The Author's Guide to Recording Your Own Audiobooks

by

Heidi Wessman Kneale

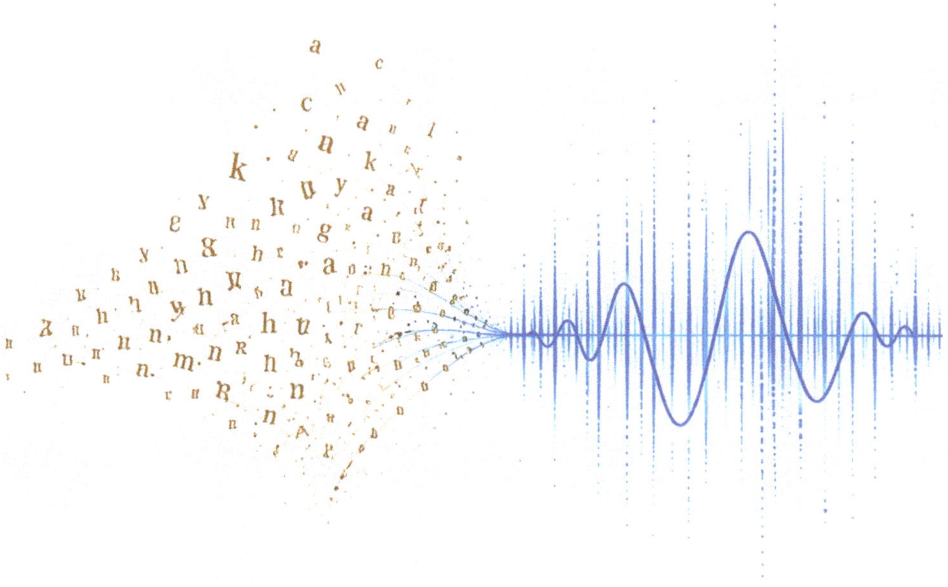

How this author did it, and you can too!

The Author's Guide to Recording Your Own Audiobooks
First Print Edition

Printed in Australia by Ingram Spark

Please note this book is written in Australian English. You may *realise* certain spellings will be *colourful*.

ISBN 978-1-923851-00-9

This publication is purely to show you how this one author with a sound engineering background records an audiobook. This is her method. Several other methods exist out there. If anything, this publication may actively discourage you from recording your own audiobook, and may point you to resources that can do the work for you. If you happen to gather useful information from her process that helps you produce a better audiobook than you would have without, kudos! May you succeed better than you hoped you would.

I want to acknowledge and pay my respects to the Whajduk and Binjareb people of the Noongar nation, the traditional owners of the lands and waters of Noongar Boodja, who for thousands of years, have maintained a tradition of oral storytelling. I'm grateful to be able to live and work on this land. I pay my respects to the elders—past, present and emerging.

Table of Contents

Welcome

If you're reading this, it means you're interested in knowing more about how audiobooks are created, and may even be considering creating your own.

Audiobooks are a lovely and convenient format for consuming the books you want to read... or rather, listen to. Many readers love audiobooks. They choose audiobooks over other formats for many reasons. Maybe they're looking for something convenient; they can listen to an audiobook while doing something else—art project, driving, going to sleep at night. Maybe the format is one that's more accessible to them, overcoming a barrier that would otherwise deny them access to literature – illiteracy, dyslexia, physical blindness and more. Audiobooks also give a sense of companionship. Here you've got a narrator reading you a story. Many people find this comforting and enjoyable. The situations that cry out for an audiobook are endless.

One thing is for sure, the format of the audiobook is here to stay.

I hope this book offers insight into how I create an audiobook. Whether you attempt to create your own, or outsource some of the process, knowing how an audiobook is created will help you make a better product.

As this manual was adapted from the RWA Academy online course, some content has been changed to better suit a general audience. Many of the external resources, such as sound files and external video links have been included as links to either my website (https://heidikneale.com.au) or my YouTube channel (https://www.youtube.com/user/hkneale) to enable you to better engage with the content.

As you read through this book, if you have any questions, you are welcome to contact me via my webpage. I'll do my best to answer them, provide clarification, or point you to further resources. Please keep in mind that I have a busy life and don't always get to my email in a timely manner.

Enjoy!

Introduction

So, you're considering the idea of turning your book into audiobook format. You didn't go straight to ACX or some other audiobook production company. Or maybe you did. Maybe you saw how expensive it can be to create an audiobook and this daunts you. You're thinking, "Maybe I could record my own?"

That is a possibility. I've recorded my own and I've provided training to other authors who wanted to do the same. The purpose of this book is to show you how it's done, or rather, how I did it.

This is my method. Other methods exist out there-different software, different processes, different resources. This is not an exhaustive guide of all options. This is a story of how one person (me!) created audiobooks that are currently available on Audible, Spotify, and more. Recording an audiobook to the quality insisted upon by ACX, INaudio or Spotify isn't a basic task. There is a bit of theory and nuance behind it. Unlike other guides, I include information about why we do what we do when we record, edit and master an audiobook.

While I am supportive of audiobook projects "narrated by the author", I realise not everyone is in a position to do that. I include several "homework" assignments so you can investigate, in a safe space, what it's like to record an audiobook. This taster will help you determine if recording an audiobook is for you. If you find one aspect (narrating, editing, etc) might not be your cup of tea, I provide information that can help you source alternatives.

Audiobook production doesn't have to be an all-or-nothing proposition. Maybe you're okay with recording your own raw soundtrack, but outsourcing the editing. Or maybe vice-versa! Maybe you don't have anywhere at home to turn into a sound booth, but your local library has space you can use for cheap or free.

Whatever your obstacle, I hope this book shows you ways that will help move you closer to turning your works into beautiful, commercial audiobooks, especially if you live in Australia or New Zealand.

Reasons for wanting to record your own audiobook:

- You can't afford to pay for someone else to do it.
- You may have the skills and resources to do it.
- You haven't fully thought it through, but hey, why not?

And these are good, valid reasons.

But just because you want to do something doesn't mean you will be successful at it. Now, I'm not normally one to discourage someone from giving something a go–quite the opposite! I'd love for you to give it a go! I want you to go through this book, try out the exercises, make a small project, and see for yourself what it's like to record an audiobook. Maybe you'll be good at it, possibly brilliant.

And maybe you'll see the process for what it is. Long, fiddly and surprisingly subtle. Me, I love it. I could sit in my studio and record for hours. I could recline in my office and edit for days. I could spent who-knows-how-long nuttering at my piano and my DAW to draft the right kind of soundtrack.

And maybe you could too.

Or probably not.

You won't know until you try.

Once you do, if you find that maybe audiobook recording isn't for you, at least you'll have a good understanding and appreciation of the process. (This might help take a little of the sting out, should you choose to fork over your hard-earned for a passion project.)

Reasons NOT to make your own audiobook:

- Your voice is terrible. You sound awful, you sound flat, you sound weak... whatever.
- You don't have a good ear.
- You are terrible at editing.
- You don't understand how to use the software correctly.
- You don't have an adequate space for recording.
- You don't have the patience for a long, tedious, and lonely editing process.
- If the thought of sitting alone in a dead-quiet space for hours on end with nothing more than the sound of your voice to keep you company, maybe recording and/or editing isn't for you.

The good news is, this isn't a "Win All or Lose All" situation. The audiobook-creating process has several skills and steps. You don't have to do it all. You are welcome to outsource certain bits of the process.

If so, it doesn't matter. You're not a failure if you have to share the burden of creating an audiobook.

Money tends to be the biggest reason many authors consider doing it themselves. Audiobook production is long, involved and requires talent and skill. No wonder it costs so much money!

And yet...

And yet.

With the onset of technology in the past decade, what was once an exclusive process is now available to anyone.

When I got started in electronic acoustics, I had access to a university sound studio sequestered in the basement of an old chapel the university had acquired, surrounded by tens of thousands of dollars of elite analogue equipment. Our cutting-edge technology was a Mac Classic computer. Our microphones cost thousands of dollars and we plugged everything to everything else through thick cables. I had to fight other students for studio time.

Today, with a decent laptop, a couple-hundred dollar USB microphone, some free software and a closet full of fast-fashion, a determined author can create quality audiobooks just as good as anything you find on Audible.

I love how far technology has come, how it gives you and me the opportunity to create something wonderful without needing to splurge on expensive equipment. Maybe you already have everything you need. Maybe you're ready to give this thing a fair go.

Please do.

Please dive in to this book. I strongly encourage you to attempt the homework I've assigned and get a feel for the process.

At worst, you'll understand what goes into creating an audiobook. At best, you may upskill and find great satisfaction in this, the creating of something new.

Good luck!

What about other audio projects?

Good question. The techniques I use in this book to record, edit and master voice audio can be applied to projects other than audiobooks. Maybe you want to do a podcast. Maybe you want to improve the sound on your long-form YouTube videos. The skills in this book will help with those projects as well.

However, this book focuses on how I create an audiobook and addresses the challenges that came with such an undertaking. Feel free to take away from this book what you want. I will make no guarantees regarding your end product. You do you, boo.

Multimedia Disclaimer

This book contains links to online audio files I've created, web sites (mine and others), and YouTube videos either I or others have created and posted freely to YouTube. These links are valid as of the date of publication. Since I do not have control over all the sources referenced in this book, there's a chance some resources may get outdated, updated or removed at a future date. While all effort has been made to present useful information in a print format, I cannot guarantee all referenced material may be available in the future. Apologies.

If you do come across a defunct reference in this book, you are welcome to contact me via the contact form on my web page (https://heidikneale.com.au) and let me know. I'll do my best to update future editions as availability and resources will allow.

Chapter 1 - A Good Mic and a Good Voice

When it comes to creating an audiobook, the most important requirements are a good microphone and a good voice. Yes, you need a good space in which to record and some sort of software in which to edit and some reliable aggregator or distributor to get your audiobook out there, but a good audiobook first begins with a good microphone and a good voice. If you've got these, you're off to a good start.

Welcome to the first steps in learning how I create an audiobook. May you have a great time learning about audiobooks and following my process. As part of this, I have some fun homework for you to try out to give you a taste of creating a sample audiobook. I hope you attempt the homework and create a satisfactory small audiobook project.

That's right. I want you to give the process a go. I'm not going to tell you it's easy and that anyone can do it out of the box, but I will break down all the steps I use to create an audiobook, show you how with online, downloadable examples, give you alternatives, and provide you with sufficient information that you can take away and improve your chances of creating a good audiobook.

Getting Started

I prefer and recommend the editing software **Audacity.** Download and install this software to your computer from the Audacity Team website: https://www.audacityteam.org/download/

Available for PC, Mac and Linux.

If you already have another kind of audio software like Adobe Audition, that's fine too. Most of the theory I'm presenting will apply cross-platform. *Don't go* ***buying*** *anything new.* Audacity is freeware and I have found it to be just as good as Adobe Audition and others when it comes to recording audiobooks, plus it has some really nifty plug-ins specifically designed for audiobook production.

(If you have access to Adobe Creative Cloud suite through work or school or some other method, Adobe Audition comes with it.) You may have another audio program like GarageBand or maybe even a DAW. I can't say I've really used these other programs extensively as audiobook-makers to be able to give good advice, but will not discourage you to try them. Please keep in mind that the steps I show here are specifically for Audacity. However, the theory behind engineering sound is pretty much the same regardless of which application you're using. Please adjust the steps for your particular software as needed.

But really, just grab yourself a copy of Audacity, even if you have other, fancier software. It's simple to use.

First thing to consider is **Your Project**. Do you have a short story or a chapter from your book in mind? If not, now's the time to raid your files for something you'd like to play with. I recommend something about 3000 words or so (give or take a thousand). Also think about your purpose for recording this project. Are you looking for a marketing giveaway? A free gift for those who sign up for your newsletter? Or are you thinking of narrating the first chapter of your own full-length books for professional reasons? For this project, don't choose anything too big. Choose it and hold onto it until chapter 4.1 (p129).

Before we dive into your project, I have several homework assignments designed to give you a simple, guided practice of skills that will help you be a better audiobook narrator. You'll find them scattered about the book. I have a summary list of homework at the back of the book.

Next, we'll have a look at **hardware**. This also includes the working space. I'll show you my setup. You might be surprised at how simple some of my equipment is. I'll explain the purpose of my equipment and why you want to splurge on some things and for other stuff, whatever you've got to hand will do.

Essentially, I'll show you what works for me and why, and you can consider what you can make work for you with what you've got.

If you just want a small promo project or are just trying things out, chances are the equipment you have now may suffice. But if you are thinking of recording an audiobook for professional distribution (like on Audible or Spotify), you may wish to invest a hundred dollars or more on a good condenser microphone.

For your first project, I STRONGLY recommend you **start small**. DO NOT under ANY circumstances, choose a whole book project as your first project. Therein lies madness. I want you to do something small, that you can complete in a short period of time, to be able to fully experience the entire process of audiobook creation without hitting The Wall and quitting a few chapters in.

Final thing for Chapter 1 is we'll be looking at some **Voice Techniques**. Being able to give a good read is vital for an audiobook narrator. We've got to lay down the best raw track we can. All the editing and post-production in the world will not save a poorly-laid down track. We want to sound good, whether on tape or live. Our voices are important tools, as important as the words we read, for conveying beautiful stories. Part of our chapter on Narrating (chapter 2) will involve you recording a few minutes of yourself reading and some self-analysis.

1.1 Hardware Introduction

Hardware is the point of contact between my voice and the recording. I wanted hardware that was suitable to my purposes but wasn't going to be too expensive. I lucked out in finding an excellent entry-level condenser microphone at a price I could afford. Everything else I happened to have at home.

For your setup, I don't recommend you attempt to precisely duplicate what I've got. Most of my equipment I acquired over a period of time for the purpose of engineering sound. For your needs, chances are you've got most, if not all, of what you need to create a simple recording. The only thing I might recommend is that, should you wish to go pro, you get a good mic. If your aim for this project is simply a marketing sample, then the equipment you have might suffice. But if you discover you want to pull off a full-length novel as audiobook, definitely level up in the microphone department.

Here's what I've got:

1. Microphone. This is the most important part of my equipment. This is the device that will capture my sound waves and convert them into a digital format that I can save, edit and play back. If you are going to spend money, this is where you'd spend it. You do not want to go with a poor-quality or too-cheap microphone.

I have an SE Electronic X1A Studio Condenser Microphone. I love it! I've got some really good results with narration with this mic. I got it brand-new in 2014 for under AU$150. I stick it on a microphone stand with a boom arm so it swings in and sits right in my face.

This mic doesn't come with a built-in pre-amp, so I need to plug it into my...

2. Pre-amplifier. Most microphones, especially condensers, need to be connected with an amplifier of some sort to provide power and boost their signal. Some modern off-the-shelf microphones, such as a Blue Yeti, come with a pre-amp built in and can be plugged directly into the computer. Some computer sound cards come with pre-amps built in, and this might work for some kinds of microphone, but I don't know if I'd trust it to fully boost my condenser mic. It really sucks a lot of juice to make it go.

I have plugged my X1 mic into a Behringer Tube UltraGain Mic100 vacuum tube pre-amplifier with limiter because it needs the extra power. Why this model? Because we're sound nerds at the House of Kneale and happened to have one sitting around. The thing's like fifteen years old, but it still does what I need it to do. If you're looking for a separate pre-amp, you can generally find an adequate one, brand new, for about AU$100 or less. Due to its age, my pre-amp is an analogue one, so I've got it plugged into the analogue sound port on my...

3. Laptop. Yep, I use a bog-standard laptop to capture and edit my sound. I work on a Dell Inspiron 3148. It's about six or seven years old but does what I need it to do. I think it cost me about AU$400 new at the time.

Laptop Specs

- Intel(R) Core(TM) i3-4010U CPU @ 1.70GHz, 4.00 GB RAM
- Windows 10 Home 64bit
- Realtek High Definition Audio

I've also got a desktop computer I could also use, but for actual sound capture, I prefer my laptop, because...

- It's portable. I can take it into my studio or wherever I'm recording.
- It's quiet. The hard drive is solid state, so no clicks and pops while it's reading/writing. Also, no fan noise.
- It has a touch screen, so I can avoid the clickety-clack of a keyboard when I'm doing stuff while recording.

4. Environment. Where you record is as important as what you're recording with. Our environment can greatly affect the sound quality of our files.

I am fortunate enough that my husband's hobby is hard-core sound engineering. When we built our house about fifteen years ago, he wanted a proper Home Theatre. None of this converted-lounge-room crap for him. He wanted to control all aspects of light and sound. Result: I have a beautiful home theatre with thick carpet, curtained walls and insulated sound baffles in all corners. Echo is all but non-existent in this room. This is where I record.

Not everyone is so lucky to have such a space in which to record. But fret not. I guarantee you do have the space and/or equipment at hand to create an adequate recording spot with minimal unwanted sound. Walk-in closets are imminently ideal. The confined space controls the bouncing of sound and plenty of clothes are excellent when it comes to sound-baffling. If you don't have the closet space, consider a blanket fort. I kid you not, do not underestimate the value of a blanket fort.

Even in the best of sound rooms, there may be a PEBSAM issue. (Problem Exists Between Source and Microphone) These are little performance issues that the mic picks up. Breath sounds and plosive pops cause these. This is easily solved with a Pop Filter. Now, you can go out and buy one for AU$10-20 or you can make one at home out of an embroidery frame and some nylon stockings, among other materials. Feel free to google.

Mine is a nylon fabric frisbee I picked up for free at the Perth Royal Show from the Murdoch University booth as a promo item for its Environmental Science degree. Not that I need a pop filter too often, as my mic is very good at not recording pops.

We'll talk more about environment and how to create your good recording space in a later chapter.

5. Earphones or speakers. You'll want good quality ones for the editing stage. Tinny little cheapies aren't going to do you any favours. But if that's what you've got, that's what you've got.

I don't listen to my recording while I'm making it. There's no need, and I find it hinders more than it helps. I'm only laying down a single track and I don't have to worry about blending or syncing with another track. Essentially, I'll record a scene, then I'll listen to it for potential issues that mean I need to re-record. It's rare that I need to re-record, but it's not unknown. Once had to redo a recording session with three scenes of a chapter because I had line noise from an improperly-plugged in cable.

When I am reviewing and editing, I listen on some really nice Audio-Technica ATH-MSR7 over-ear hi-res earphones. They're rather high-end for my needs.

Hardware Issues

When we have the right hardware, we're going to lay down some good sound. But if a piece of hardware isn't quite up to scratch, we'll have some issues.

1. **Microphone issues**; with a cheap or improper model you may get issues like:

- frequency dropout, usually in the bass end, leading to a tinny sound
- variable amplitude, where the sound gets louder or softer randomly
- low amplitude, where it doesn't have enough power to drive the sound (thus, the need for a pre-amp)
- extra noise, such as line hiss/buzz or a general insensitivity that catches all background noises.
- lack of range. If you step back from the mic you risk the sound dropping significantly.

2. **Computer issues**.

- Not enough memory. When you're recording or editing an audiobook, you work scene by scene. It's not uncommon for these files to be a good thirty minutes in length or more. Raw audio files can take up a lot of memory. If a computer can't handle it, it can slow down or even crash.

- Noisiness. Computer fans are the biggest culprit when it comes to recording noise. Their cooling fans can be quite loud. Magnetic hard drives can also be noisy, ticking as they read/write. The keyboard and mouse can also make noise when you click them. Occasionally computers will talk back to you, in pings and notifications and other "useful" noises.

3. **Environmental noise issues**.

- I have a podcast I listen to on a regular basis, but the podcaster, for some unfathomable reason, did not think to remove his noisy budgerigar from the room. You'd think he would have thought of that. Some noise we can solve. Some is harder to control.
- One noise we can't prevent is line noise. This is the buzz of electricity through the cables (lines) that connect our microphone to a computer. Your software will pick this up; it'll show up in the noise floor. The more equipment we have between our mic and our computer will contribute to the line noise. Most of the time it is low, or it can be controlled through gain (volume). This noise is constant, but that can be a blessing, as a constant, regular noise can easily be filtered out with a Noise Reduction pass in post-production. I'll cover that in the Editing Chapter.

Up for some easy homework?

Find a spot in your house. Sit down, get comfy, close your eyes, and listen.

What do you hear? What's the background noise?

Right now in my house I can hear:

- Fan of a computer
- Click of a hard drive
- Refrigerator hum
- Traffic from the road
- Train
- Neighbours
- Birds
- Barking dogs (neighbour's)
- Snoring cat (mine)
- Daughter typing away in the other side of the house

Most of the time I ignore these noises. But on a sound recording, they'll stick out. It's good to develop an awareness of the ambient noise in your environment.

When the world isn't busy making noise around you, the "silence" you hear is called Room Tone. Nothing wrong with this. It's Life. It's the Universe. It'll be your noise floor of your recording. It's not necessarily something you want to eliminate entirely, as it adds a sense of "space". However, you want it minimise it so it becomes unnoticeable, not necessarily un-present.

Homework, part deux (hands on)

Got a mic? Got a computer? Got recording software? Or do you just have a phone? Whatever you've got, click record and capture about thirty seconds of ambient sound.

Next, plug in the best set of earphones you've got and listen to that ambient sound. You may need to turn up the volume of the recording to hear what's really in that background.

Does it sound the same as when you listened live? Did you hear anything in there that you didn't hear live, like line noise?

Microphones

A microphone isn't just a microphone. They are beautiful pieces of engineering designed to capture different types of sound. Depending on what kinds of sound and from how big a capture area determines which sort of mic would be best for your needs. They vary widely in the types of frequencies they capture, the size and range of area they capture, and more.

Three of the most common types of mic are:

- Condenser Mics
- Dynamic Mics
- Ribbon Mics

These mics come in different flavours for different uses: diaphragm size, polar patterns, etc.

For recording a narrating voice at home, you will most likely want a **condenser microphone**. They're best for controlled environment with no extra noise and optimal acoustics, meaning sound won't be bouncing around off your walls and back into the mics. They're extremely sensitive and will pick up the beautiful nuances of the human voice. The downside is they will pick up extra noise. But then, you'll be constructing some sort of recording booth from your walk-in closet, so this shouldn't be too much of an issue.

A condenser mic will need a pre-amp to power it (providing something called 'phantom power'). Some mics like a Blue Yeti USB mic come with a built-in pre-amp, whereas others like my SE-X1A need an external pre-amp.

Other mic types:

A **dynamic microphone** is the kind you tend to see on stage at concerts or other outside venues. They're really good for loud situations (like concerts) or times when you want to capture a wide range of volume (newscaster interviewing outside). One benefit of dynamic mics is that they have passive electronics and don't need phantom power like a condenser.

Can you use a dynamic mic to record an audiobook? Yes. If you happen to have one of good quality lying about, and the ability to connect it to your laptop, go right ahead. A dynamic mic with a cardioid polar pattern would be suitable to this kind of solo voice work. Being less sensitive than a condenser mic, this might work out better for you in a studio space that might have a bit of bounce or echo to it, as they will be less likely to pick up random noise.

A **Ribbon microphone**. These are rare, fragile and expensive. Unless you're a total sound nerd, you won't have one of these, nor would you want one of these, especially when a good condenser will suit your purposes just fine. (But oh, they sound so sweet and warm!)

So, what would a newbie like you need, microphone-wise? For your first project, a simple USB condenser mic will suffice. They come with onboard pre-amp, which is what you want. If you choose well, a good USB mic can capture the human voice sufficiently for an audiobook or podcast. Once you've done a small project or two and decide you like this audiobook schtick, you can up your mic game later.

Some brands with good reputations for USB mics include Audio-Technica, Blue Yeti, Rode, Sampson and Shure. Be prepared to pay between AUD$100-200 or more for one of these.

Confession: I've not used any of these for the actual recording of an audiobook. While I've only used a few of them for other purposes (and had no issues), I do not have experience in audiobook production with them. However, I do know several other people who have used them, especially Rode and Blue Yeti, who report good results.

Please do not scrimp on the quality of your mic. While you might find a cheap USB condenser mic for $39 and think it a great deal, if you plan on recording an audiobook for commercial release on Audible or Spotify, these cheap mics might not deliver the strong, original quality you will need to pass Quality Assurance. Remember, while you can make a good signal sound great, you cannot make a bad signal sound good.

While you may have picked up this book hoping to reduce the costs of producing an audiobook as low as you can get it, absolutely do not cheat yourself out of a sufficiently-good mic. The little head mic you used for Zoom meetings in 2020 will not suffice. They're just too cheap.

An audiobook is meant to be warm and subtle and personal. A good mic will capture this in your performance and translate to a better listening experience to your fans.

Alternatives

Nervous about spending a few hundred dollarydoos on a mic if you're not sure you want to record a full audiobook project? That's fine. There are other options you can explore while you're learning the process. For the sake of the exercises in this book, yes, you can use the cheap headphone mic, or the toy-like dynamic mic that came with your copy of SingStar for Playstation. Just keep in mind that while the quality of sound from these mics isn't going to pass QA at ACX, it might be good enough for a personal project like a newsletter giveaway or an Instagram/YouTube promo feature.

Other options are available to you to try out recording. You can always find a local studio and rent out studio space for a half-hour or so. Some public libraries offer studio space for its library patrons. My local library in Western Australia (as so many of them do) offers this service. I absolutely recommend investigating this option. Having access to a cheap or free recording space for capturing raw sound that you can take home and edit on your laptop is a great alternative to sourcing your own mic or creating a recording booth at home.

One final word of advice: I know the "Mic Drop" is quite The Statement, but in reality, never, ever drop your mic. They are sensitive.
They do not like to be abused, dropped, shouted at too loudly or insulted regarding their dress sense. Treat them gently, treat them well, and they will serve you to the best of their ability for years to come.

Recording Space

I am fortunate in that I have an over-engineered Home Theatre that doubles well as a recording studio. This theatre/studio space features thick carpet on the floor, full-length blackout curtains on two walls, acoustic tiles on key surfaces, and insulated sound baffles in the corners. While it isn't a total dead room, it is padded sufficiently to reduce any sound bounce a hard, flat wall can create.

Unless you live with an audiophile sound engineer-by-hobby like I do, chances are you don't have a pre-existing space like this.

But you can create one.

The secret is to choose a space where you can easily and cheaply create a more soundproof environment. It doesn't have to be big at all, just big enough to fit you and your mic comfortably. A walk-in closet is ideal. These small spaces tend to be carpeted on the floor and full of clothes, which make an excellent sound baffle. If there's room in there for your equipment, including a comfy chair (if you choose to sit), very little change is needed to make it a workable space.

Another alternative is the good old cardboard box sound booth. If you don't have a dedicated space like a closet, or are short on any kind of space, I've seen a little recording booth made out of large cardboard box. Plenty of YouTube videos and online guides for making one.

Essentially, you get a large enough cardboard box, line it with soundproofing (like the egg carton wall dampening insulation) or a thick quilt, and you're good to go.

I've never made one, I've never used one, but if that's all you've got, it's worth having a look at what others have done.

One easy way of improving soundproofing is the good old blanket fort. Seriously. A nice, thick doona or other sufficiently bulky blanket propped over your recording area can dull down the sound of a space. This is the poor man's version of the cardboard box sound booth I mentioned earlier. I recommend getting online on YouTube and having a look at how others have created their own blanket fort recording booths.

More Homework

So, you've got your recording space? Good. Set up your mic and your laptop, and spend a minute or two recording your ambient sound. What's the sound floor like? In an ideal space, it should be low, preferably below -65dB. Next, do a voice check. Record yourself narrating a few paragraphs, then listen to the results. Are you getting any echo off the ceiling? Any background noise or other unwanted effects?

If you get any echo or other unwanted noise due to room acoustics, you'll need to soundproof more.

1.2 Narrating – Voice Techniques

Good voice techniques will help you deliver a better performance. The following info is useful if you're:

- Recording an audiobook
- Doing a podcast, either as a host or a guest
- Creating a YouTube video
- Speaking in real life, such as on a panel at a convention or at a library event

---

Homework Part I

BEFORE you read through this section:

Get your recording setup and record a good couple of minutes of you reading something. I don't care what you read. Just record it as-is. Save it, as we'll be looking at it later.

Got your recording? Good. Now, go ahead and read through this section.

The Importance of Good Technique

Why is good technique important? When I start recording audio, I want to produce the best quality raw file I possibly can. This means my voice acting technique needs to be good. Silk purses, sows' ears and all that. The better your initial recording, the better your final product. You can make a good recording great, but you can't make a bad recording good.

While editing a file can clean up breath noises, sneezes, mistakes and other unwanted noise, it can't fix the noise in the sound I do want nor can it add what's not there.

Good performance is vital to a clean track. The best quality final product happens when I start with a clean track. Sure, there's lots of editing tools in Audacity I can use to process this and alter that, but overprocessing a track will degrade its quality. (I'll go more in depth into this topic in Chapters 2.3 and 4.1. The less processing you need to do on a track, the purer the final product and the more likely I'll be to pass QA if you're going pro.

Here's some voice techniques I use when I'm performing. By keeping these things in mind, I'm more likely to create a clean, raw track.

1. **SLOW DOWN**. One of the biggest errors an inexperienced narrator can make is speaking too fast. Until you gain more performance experience, your articulation will suffer under greater speeds. You don't want to muddy your words together. I like to slow down. This gives me space for tempo variance, depending on the drama of the scene I'm reading. Urgent scenes I'll give a quicker speed, whereas something not so urgent will get a slower speed. I aim to make my words clearer and easier to understand.

Good articulation is preferred over a faster pace. On average, a slow narration speed will be about 100wpm (words per minute), a medium speed is about 150wpm (my preferred comfort range), and a fast speed is about 200wpm. When starting out, there's nothing wrong with aiming for a slower speed. Accuracy and speed will come naturally with experience.

Slowing down will also reduce issues such as loud breath noises.

2. **Articulation**. I want my consonants smooth and my vowels clear. By taking a bit more time to slow down and giving good articulation to my words means a clearer, more open sound. After all, you don't want your words to slur together or get muddied. This means I need to ensure consonants are solid and vowels are distinct.

Most audiobook listeners adjust the speed of playback to suit them. Clarity of my words will serve my audience better, should they choose to listen at 1.25x, 1.5x or even 2.0x speeds.

3. **Voice positioning**. Where I place my voice within my head makes a big difference. Ever hear someone with a gravelly, throaty voice? How about a nasally whine? (And don't get me started on vocal fry!) Imagine listening to that voice for hours. I want to give my listeners a pleasant experience for time we spend together. I want my voice to soothe, to carry them, to be pleasant to the ear. Positioning my voice so it sounds like it's resonating from the front of my mouth gives me the best sound. More in Chapter 3.4 including a video link demonstrating vocal exercises and voice positioning.

A benefit of good voice positioning is your voice will last longer. For me, good voice positioning means I can spend hours recording without going hoarse (laryngital fatigue).

4. **Vocal fillers**. This is known as speech disfluency. These are words like "Umm… Uh… well… Hmm…" that we use to fill in the spaces between our phrases. We all do them, sometimes to buy us some thought time and sometimes to give a sense of continuation during conversation so listeners don't assume we're finished speaking. As my audiobook

audience is a captive one, I don't have to worry about keeping their attention. Also, as my narration script is sitting right before me, I don't have to think about what I'm going to say next; I just read the lines. Smooth, pure, and without any extra sounds. I let the words shine.

The problem with vocal fillers is that most of the time we're unaware we're doing them. They can creep in during narration when we're not paying attention, especially if we're pausing between sentences, paragraphs or page turns. Awareness of vocal fillers helps to eliminate them from our speech. Sometimes they can be edited out in post-production, if they slip in, but it's best to get used to not inserting them in the first place.

> *Do not be afraid of silence. Silence serves you well.*

5. **Plosives, pops, clicks and other strange mouth noises**. Yep, these happen. Again, if you're not used to listening to your voice, you might not realise you're doing them.

A plosive is a consonant like P or B that gets pushed out with air. It's a bad habit most English speakers pick up. French speakers don't have this issue because they soften the impact of their plosives. I had to practice my plosives in a French Immersion class at university by pronouncing P words into a tissue held up before my face. If the tissue moved, I was pronouncing it too hard.

Sibilance is the sound of S. Try words like: duchess, hiss, kiss, susurrus... you get the picture. S sounds can be extra-harsh if you push them too hard with breath. Sometimes sibilance is caused by drawing too deep a breath too quickly. The inward hiss can be intrusive.

Pops and clicks can come from non-word sounds, or an extra harsh air push on certain consonants.

While filters do exist to de-click and de-ess, this is considered an extra layer of processing. Too much processing = bad. Best to avoid this by laying down the best quality track you can.

Strange mouth noises. The mouth is a wet, gross place. When it's less wet, it's even grosser, because it gets sticky. When that happens, "wet" mouth noises are made. They're sticky and sound like the snapping of chewing gum and are most unpleasant to hear. It sounds like that really gross kid in high school that, to think of kissing them, would make you shudder. A plentiful bottle of water, less facial tension and awareness usually clears this up.

Personal sounds. Sneezing, coughing, clearing your throat and yelling at the cat. Not much you can do about these, but don't worry too much, unless you're performing live, when they can jar your mojo. When you're making personal sounds, you're not actually narrating, and most of them can be cleaned up with an editing pass.

The good news is, you can train yourself out of bad plosive habits. Also, using a pop filter can ease up on some of those breathy bad habits.

6. **Inflection**. Nothing is more boring than a flat-pitched voice. I once attended a convention where the author Guest of Honour gave an hour-long reading from one of his novels. Man, did he drone on. More than one audience member drifted off. It was awful. Pitch variation and tone patterns can add interest and variety to a narration. Here's some simple inflection techniques:

- At the end of a question, pitch should rise up.
- At the end of a statement, pitch can drop.

- A pitch rise can emphasise a word, indicate surprise or indicate lightness.
- A pitch drop can add gravity or disappointment.
- Don't forget to include pauses in your inflection. Beats and rhythm also play a vital part in a good performance.

Homework Part II: How's your reading voice?

Go back to that recording you did before reading through all this. Listen to it with a critical ear. What did you notice about your reading before? How's your pitch/tone? How's your reading speed? What else did you notice about your performance? Is there anything you'd like to change? Write this down. Keep all this in mind while you proceed to this:

Homework Part III

Using whatever equipment you have access to (into your mobile phone or into Audacity—doesn't matter for this assignment), read a scene from one of your novels. Set a timer and record five minutes of you narrating. When that timer goes ding, stop.

Note where you stopped in your reading and run a quick word count. Divide this number by five to get an approximate words-per-minute (wpm) speed.

Then go back and listen to the whole recording.

- What do you notice about your voice?
- Do you like the sound of your voice?
- What forms your opinion?
- Were your words clear and easy to understand?
- Did you notice any plosives, sibilance or other rough edges?

- Any non-verbal noises you noticed?
- Did your voice sound engaging or was it flat?
- What would you like to change or improve in your voice?

Feel free to re-record again. Practice and reflection will help your voice improve.

Reasons why you shouldn't narrate:

- You can't get your voice to work well
- You don't like how you sound
- You're a terrible actor
- You can't narrate for more than half-hour without going hoarse or getting tired.

That's fine. Vocal muscles are muscles. If you're not used to using them for long periods of time, it will be difficult.

But I'm glad you gave it a try.

If narration is proving difficult because you're not used to it, you have two options:

1. Build up your muscle strength with vocal exercises
2. Outsource the narration. (more on this in Chapter 5)

1.3 Speaking Speed - Words Per Minute

On my website you will find an .mp3 sound file of different speaking paces. This will give you an idea of what different speeds sound like.

Download sound file: Words Per Minute (see page 182 for link, or visit https://heidikneale.com.au/AGRYOA/)

Access and listen to the sound file. You can follow along with this text:

Text:

100wpm: This is the pace of narration spoken at one hundred words per minute. For everyday dialogue, this would be considered rather slow indeed, but for audiobook narration, it isn't too slow a pace. This allows the narrator to form their words well and let the beauty of the sentences shine through. Slow speech also lends gravity to a piece.

130wpm: This is the pace of narration spoken at one hundred thirty words per minute. As you can see, it's a bit faster-paced than one hundred wpm, but still allowing space between the phrases. Nobody would fault you for reading at this pace, unless it works against the text. Some text works well with a slow pace, but other text demands faster.

160wpm: This is the pace of narration spoken at one hundred sixty words per minute. This is the average speed of a normal conversation yet is still clear enough to hear the enunciation of each word without them getting rushed or muddied up. This is a comfortable pace and would serve many narrators well.

190wpm This is the pace of narration spoken at one hundred ninety words per minute. Fast speed can give a sense of urgency, when used well. Unless you're reading a fast-paced book, it's better to slow down from this speed or you will feel rushed.

Now, this is not to say that when we narrate, we must maintain a constant pace. In fact, varying your pace will serve you well. I use a faster speed to give a sense of urgency. You know, punchy sentences. Go, go, go. Get those words out and let people know that stuff is going down.

Likewise, a slower pace will lend emphasis to an idea. Gives you time to think, to process. Let your audience get used to a new idea. My speed will, in part, be dictated by the content of the text. Is it pacing that's dominant, or is it the ideas I'm communicating? What's the emotional weight of the passage? Do I need to slow down for sadness, for romance, or do I need to speed up for anger or for passion?

I prefer a comfortable, moderate speaking pace.

1.4 About your project...

I'm trusting that you're reading this book because you've got a full-book project in mind. Good!

But tuck that away for now, and start with something small. I'd like you to consider a short recording of a body of text approximately 3000 words, or about a half-hour finished audiobook. Let this be your first official project. Make all your rookie mistakes on this one before you commit to something larger. Choose your 3000 words and hang onto it for the homework on page 129.

Question for yourself: What would you like to do?

What are you recording? What do you want to do with it? What are you envisioning about this project? What are your concerns or fears? What do you hope to accomplish?

Some things to think about regarding your project:

- What kind of story is it?
- What are the themes?
- What moods does it move through?
- What are your characters like? Can you envision the sound of their voices in your head?

Grab a notebook and makes some notes.

As you work your way through these chapters and attempt the various homework assignments, please record your impressions. Write down what you do right, what you like, what's working well. Likewise, write down what you need to improve on, what's hard and what you may potentially consider outsourcing.

I recommend a notebook to chart your progress and make notes. This will help you as you learn the process of creating an audiobook.

Now gird up your loins, set up your sound booth and fire up your microphone. You've got some work to do.

Reasons not to:

Absolutely none at all! If you were willing to pick up this book, you're willing to give this small project a go, even if only to prove to yourself that you don't have to do it all. Or maybe you will. You will never know that until you make something. It doesn't have to be perfect. Doesn't even have to be good. But it has to **be**. Let's have some fun with it!

Give yourself permission to make mistakes.

Chapter 2 - Recording and Editing Sound

How We Record Sound

By now you've had a chance to settle in. During Chapter 1, I hope you had an opportunity to check out your available hardware and software, and have had a chance to listen to ambient sounds and think about what kind of recording environments are available to you. You've also thought about your voice quality and the speeds at which you speak, and maybe even recorded a few test samples.

In this chapter we're going to have a look at some sound wave theory. This will help you understand what you're looking at when you're staring at those pretty blue shapes you've just recorded in Audacity, and when we're editing those pretty blue shapes.

We'll do some hands-on with Audacity. I'll show you what I do when I record a track, including how I tweak my sound meter and Audacity's levels for capturing the best sound I can. I'll also provide a raw .WAV soundtrack you can download from my website open in Audacity and have a play around. (Or you can record your own passage of your own choice.) This is where you can have some fun and explore all that Audacity has to offer the audiobook creator.

2.1 Sound Wave Theory

I'm going to cover a little bit about sound waves—how they're created, their anatomy, and how they're stored digitally. I've included several links in the glossary at the end of the book for those of you who would like to go more hard-core into theory.

Making Waves

When I speak, I create pressure waves in air that, at certain frequencies, are interpreted as sound. Different sounds create different levels of pressure. The changes in pressure levels is called vibration. The diaphragm of my condenser microphone picks up the different levels of vibration and translates them directly into electrical signal. This signal isn't strong, as the vibrations of sound are very subtle. A pre-amplifier amplifies, or increases, the strength of this signal in a way that helps preserve the fidelity of the original vibrations. This is sampled and stored digitally on Audacity in a format that I can edit and filter to create a desired result.

Frequency

Sound wave frequencies are measured in Hertz (Hz). This is how often the sound vibrates per second. We hear this as pitch, or how high or low a wave sounds.

If you were to take a ruler and wave it as fast as you could, you'd get a vibration of about 3-4Hz, a little too low for the human ear to pick up. But if you could wave it twenty times per second, you'd be waving it fast enough that the human ear might just barely be able to hear it as a low rumble. The faster the speed, the higher the pitch. (Humans can hear sound from 20Hz up to 20,000Hz.)

A sound wave looks like this:

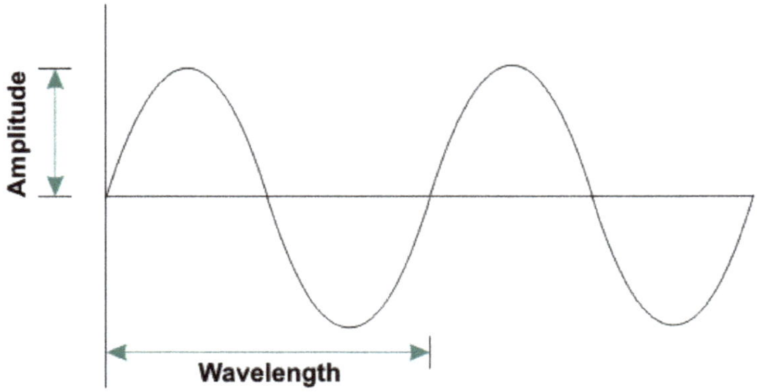

Figure 1: basic wave anatomy. A wave starts at zero, rises in amplitude and falls, and then drops below zero in amplitude, until it rises back to zero. This is a full cycle of the wave, or its wavelength. The number of waves that occur within one second is its frequency. Image credit: Wikicommons

This is a simple wave (called a sine wave). It has a wavelength and an amplitude.

The longer the wavelength, the lower the frequency, and the lower the pitch. The shorter the wavelength, the higher the pitch.

The larger the amplitude, the louder the sound. The smaller the amplitude, the quieter the sound.

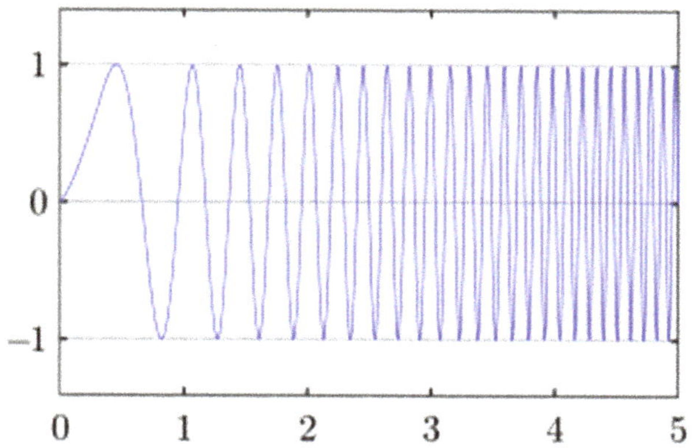

Figure 2: as time progresses, the frequency increases. Image credit: Wikicommons

In Figure 2, the frequency of the wave between 0 and 1 second is 1 Hz, as it only completes one wave cycle within a second. Between 1-2, our wave becomes 3 Hz, 2-3 is at 5Hz, and so on.

The frequency range of human hearing is from 20Hz to 20,000Hz (or 20kHz). Because you need a minimum of two samplings of amplitude to determine the amplitude and frequency of a single wave cycle, we digitally sample our sound just above 40kHz.

If you record a sound wave in Audacity and zoom in to the deepest level, you'll see the actual sampling:

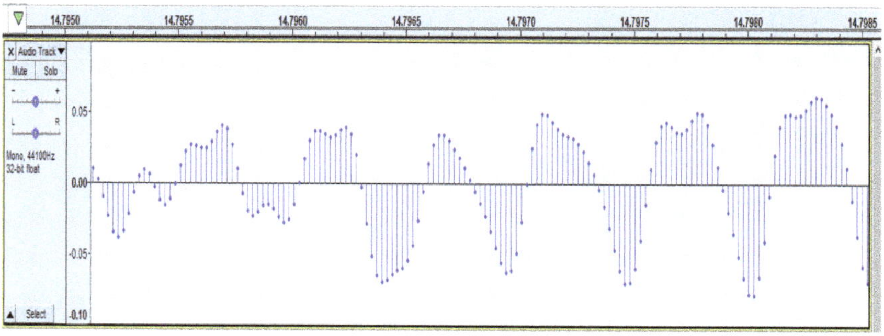

Figure 3: some sound I sampled in Audacity, zoomed in to the actual sample points.

The numbers at the top are seconds. As you can see, this sample is zoomed in to milliseconds, and shows the exact moments of sampling. On the left, it shows that we're sampling at a rate of 44100Hz, which allows us to have a high fidelity for the range of human hearing but doesn't overburden us with too much data because storage space is an issue. (You can change the sampling rate in Audacity, but for our purposes, the default 44100Hz is adequate.) If you were mathematically inclined, you could figure out the frequency of this pitch if you wanted to. How many wave cycles are there per second?

Bonus info: Narrative voice pitch

Human speaking range tends to be between 150-500Hz. (Singing pitches can extend beyond this.) The note known as Middle C vibrates at about 262Hz. Male voices tend towards the lower range and female voices the upper, due to the size of the vocal cords. The bigger the vocal cords in your throat, the lower your frequencies when speaking. That's why women tend to be sopranos and men basses.

Humans tend to speak in the middle or towards the bottom of their frequency ranges. The pitch range in which you speak is called the modal range. It's the range of notes that's most comfortable for you, and which you have the most power. For most voices, the pitch range used for speaking tends to be less than an octave.

In music, an octave is the range of eight notes: Do Re Me Fa So La Ti Do. From Do to Do is one octave. In mathematics, an octave is the doubling of the frequency. Therefore, A3, or the "La" note below Middle C is 220Hz, whereas the A4, or the "La" note above Middle C, vibrates at 440Hz.

TUNING THE ORCHESTRA (video)
https://youtu.be/npEz3hOh8CA

The A4 (440Hz) note is called "Concert A", and is the note used to tune an orchestra. An oboe (always an oboe) will play the note so everyone can hear it. How does the oboist know she has the correct note? The oboist has a sound meter and will tune her instrument to 440Hz. She plays the Concert A until the Concert Master gives the signal for the instruments to tune.

Amplitude

Amplitude is how much power a wave has.

For our purposes, amplitude will be measured in decibels (dB). Decibels measure sound pressure, which translates to us as "loudness". The stronger the pressure, the higher the decibels and the louder the sound. Decibels are a logarithmic scale.

In sound recording, decibel measurements start at a maximum of 0, which is the loudest possible sound the software and hardware can handle. (Anything over this will get clipped and distorted.) The scale proceeds *downward* into a negative scale, depending on how quiet our signal gets. -3dB is pretty loud, and -30dB is much quieter. -70dB is barely audible.

When we record, we're laying down sound that could reach up to the maximum, but we don't actually want to get there. Once we hit the maximum, any further signal gets lost, or clipped. To our ears, this will sound like distortion, and we don't want that.

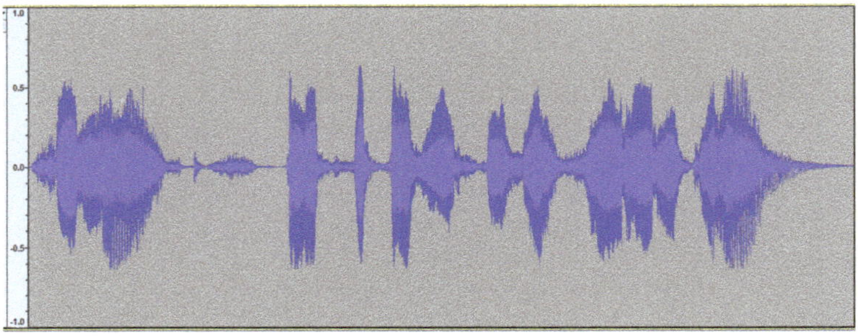

Figure 4: variations in amplitude. The bigger blue waves are where the sound is louder. At the end of this sample, you can see where the sound tapers off as I fade out.

When we record, we want a nice, warm signal coming in, but not too strong, or we risk overdriving our signal and getting it clipped by hardware and software limitations. Likewise, we don't want it too soft, as the subtler nuances of our performance might not be captured. Between my pre-amp and Audacity, I have tools I can tweak to ensure I'm getting a warm, strong signal.

Fortunately, Audacity has a sound meter I can turn on before I start recording my voice so I can see what kind of amplitude I'm getting from the sound I'm putting into my microphone.

For my raw signal, I want to aim for a signal averaging between -30 and -12dB, with a peak no higher than -3dB. To see this in action, fire up Audacity and click on the sound meter to start measuring the sound coming in. You can play with your laptop's onboard mic or you can plug in your better mic for this exercise. Quality doesn't matter for this exercise, but quantity does.

Speak into your microphone and watch the bouncy green bar. If you speak louder, or draw closer to the mic, you'll see the green bar bounce even higher.

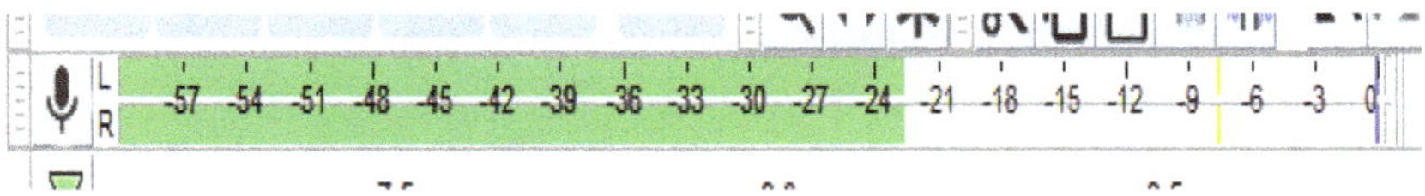

Figure 5: sound meter as I'm narrating. The green bar is the strength of the signal currently coming in. The yellow line shows the maximum peak of the signal over the past few seconds. The blue bar is the maximum signal possible. If you're hitting this, you're going too far. Really, you don't want to peak higher than -3dB. Me peaking at around -8dB in this example is okay.

Being able to see the sound meter running before you start recording will allow you to adjust your equipment so the signal you're feeding in is good without you needing to compensate quality by shouting or restraining your voice.

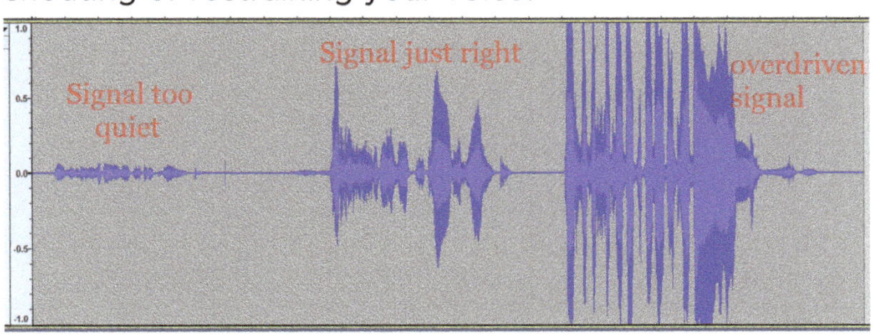

Figure 6: examples of amplitude strength while recording.

ACX (Audiobook Creation Exchange) has set the industry standard for noise levels for their Quality Assurance (QA). Don't worry too much these when first laying down your soundtrack. Just ensure you have a nice, warm signal that's not overdriven and not underperforming. We have tools in Audacity to fix any rough edges later and bring our soundtrack into compliance.

Aim for your signal to average around -23dB and -18dB. Your bar will bounce around while you're narrating, and that's okay.

Audacity lets you know if your peaks are getting too warm when the peak bar turns from green or yellow to red. If your peak bar is hitting red, or maxing out at 0dB, you're a little too warm.

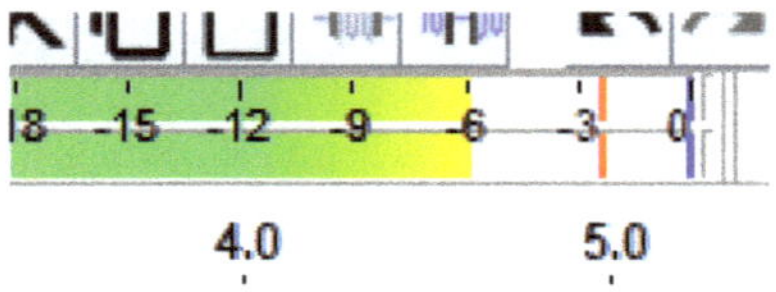

Figure 7: too warm! That red bar just above -3dB means we're driving our signal too hard and risking hitting the maximum peak.

The quiet bits between dialogue need to fall below -60dB. This silence in between our narration is called the sound floor. If this sound floor is too high, we get a background hiss that intrudes on our narration.

2.2 Introduction to Recording on Audacity

Start up Audacity on your computer and have a look at the toolbars. Yours may look slightly different to my layout, because these toolbars can be dragged around and customised. I customised mine to suit my style. All these screen shots are from Audacity 3.3.3. Later versions may look cosmetically different, but the functionality should be the same or similar.

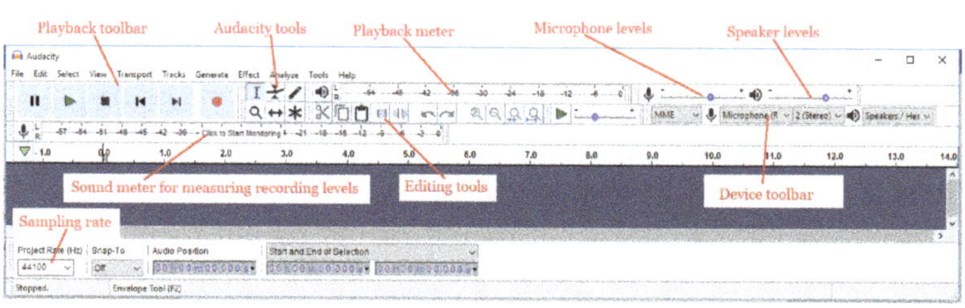

Figure 8: The toolbars of Audacity.

Audacity has:

- **Playback toolbar**: This allows you to play, stop, pause, and record sound.
- **Audacity tools**: These are the Selection tool, the envelope tool, draw tool, and more.
- **Playback meter**: this will show you how warm your recorded signal is.
- **Microphone levels**: this sets the strength of signal from your mic.

- **Speaker levels**: how loud is your output while working with Audacity
- **Sound meter**: shows how strong the incoming signal is from your mic
- **Editing tools**: includes zooming in/out, silencing, trimming and more.
- **Device toolbar**: this allows you to select your input, output, MIDI mapper and more. For example, if you're working from a laptop, chances are it has an inbuilt microphone. Since you don't want to use that one, but the one you've plugged in via your sound jack or USB port, this toolbar allows you to choose which mic Audacity will listen to.
- **Sampling rate**: how many times per second Audacity is sampling sound. 41000 is standard. This provides the most efficient use of data storage without sacrificing audio quality. Yes, you can raise this (more data, bigger files) or drop it (poorer sound fidelity), but this is the industry sweet spot.

When you first start up Audacity, you'll see an empty grey window. This is because we haven't captured any sound yet. Once we capture a sound, Audacity will automatically create a track for us.

Tweaking your hardware for the Sweet Spot

This is probably the fiddliest part of getting set up.

Get your laptop, plug in your mic, fire up Audacity, and let's get started. (Note: you may need to tweak your Sound Settings on your computer so it recognises your microphone as the input. Consult the Help files of your Operating System (Windows, iOS or Linux)

Before we record, you'll see your sound meter will have a "Click to Start Monitoring" message. If you click on the sound meter, it'll start measuring all the input that your mic is picking up. If you speak into your mic, you'll see the sound meter bounce around.

One of the uses for this sound meter is to adjust the strength of your incoming signal to ensure it's as warm as you need it. I tweak the input strength one of two ways:

- boost my pre-amp's signal coming in, and/or;
- up my microphone input through the microphone levels toolbar.

I'll often tweak both to ensure I'm getting a warm signal when I speak in my average volume voice, but not so strong that I'm getting a high sound floor. It's a balancing act.

Once I get my signal the way I like it, I can lay down a track.

Recording in Audacity

To lay down a track, click the Record button (it's the one with the circle on it) and start speaking into the microphone. Click the Stop button when done.

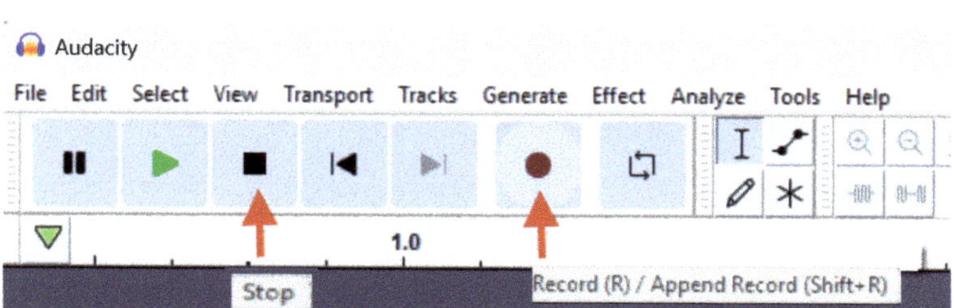

Figure 9: the playback buttons, with Record and Stop highlighted.

What's your voice like? What's the ambient background like?

Things I listen for:

- Noticeable hiss in the sound floor
- Tinniness in my voice quality
- Missing bass (lacks depth)
- Pops, clicks and sibilance
- Microphone artifacts (crackliness, line noise, etc)

These issues suggest that I need to tweak my levels, move my mic, or check my cables.

Hiss in the sound floor can mean I've boosted the signal too much. Hiss is often line noise from the electricity moving through your cables. While this can be filtered out, the filters in pretty much every sound program will also filter this frequency out of your voice, downgrading your voice quality and making it sound artificial.

So, I drop the signal on my pre-amp. Now my voice isn't quite as loud as I want. This is easily solved by moving the mic closer to my mouth. There's a desirable distance for a mic from your mouth, which is roughly one shaka away. (Hang loose, dude!)

Figure 10: One shaka away from the microphone
Photo credit: forum.audacityteam.org

What is a shaka?

It is the official hand gesture of Hawaii, representing "aloha spirit, love and local pride". https://en.wikipedia.org/wiki/Shaka_sign

---

I can move a little closer, but then I risk plosives, sibilance and strange mouth noises. A pop filter can help with that.

However, if you find yourself kissing the microphone, that's a little too close and you will get distortion because the power of your breath will overdrive your mic's diaphragm.

Essentially, it's a case of tweaking and tweaking and adjusting until you find a level you like. Once you do that, MAKE A NOTE! I've taken pictures of my pre-amp and made notes of my Audacity settings so I can easily recreate the ideal conditions, should something happen to adjust them (like little fingers twiddling knobs on your pre-amp, like what happens at my place). This includes taking a picture of me and my mic, addressing each other at the ideal distance.

There will be a sweet spot. Once you find it, congratulations!

Get your sound meter running and utter a few test sentences to see that everything is bouncing around the way you like it. When you're happy, you're ready to go.

Let's do a recording.

Laying down a soundtrack

On the playback toolbar, hit the Record button. Now, don't breathe. Don't move, don't do anything that will generate sound. Just let your mic pick up 3-5 seconds of room tone. We'll need this later. Then, take a breath, pause and speak. Lay down your narration.

As you speak, your soundtrack will be created. There is a delay of a second or two between when you speak and when you see it wiggle across the screen. This is normal.

When you're done with your narration, pause for a few seconds for some room tone, then hit the stop button.

Congratulations, you've laid down some soundtrack. It really is that simple. All the hard work is ensuring that your hardware setup is ideal, and your voice is warmed up and ready to perform. If you're giving a good performance into optimal equipment, you're going to get good soundtrack. This is the basis of a professional recording, and what you're aiming for.

What is room tone?

Room tone is the ambient sound of your room or your recording space. Every space has a room tone, because the universe is noisy! We ensure we have some room tone at the beginning and ending of our recording.

When we create our recording space, we want it to be as quiet as possible. Our desirable room tone should be as close to silence as possible. We don't actually want to silence out our room tone. (Audacity has tools that can do this, but we don't want to.)

Room tone has its uses:

- Noise Reduction. One of our post-production steps is to reduce the overall noise. If we have several seconds of pure, unadulterated room tone, we can sample that, and, essentially, remove the room tone, thus dropping our noise floor.
- Spacing. Quality Assurance requirements for ACX and INaudio and pretty much everyone putting out quality audiobooks require several seconds of room tone at the beginning and ending of each file. This is to give a natural-feeling separation between the scenes and chapters. If this room tone has been completely silenced out, the listener will sense that, and that can jar them out of their listening experience as they wonder if the audiobook has stopped suddenly, or something else has gone wrong. Room tone is comforting. Just don't let it be too loud.
- Establishes the "noise floor" for the listening experience. Having a non-silent room tone at the beginning and ending of each scene will improve a listener's experience. Their brain will catch the room tone, then filter it out. It's one of those things the brain will identify as Not Worth Listening To.

Now that you've got good track, save it.

Saving your work

In Audacity, you save your work as a project. When you first create a project, it will create a file .aup or .aup3 (Audacity Project) and it will create an accompanying data folder.

Keep in mind that sound files, especially in their raw form, can be quite large (remember that 44100Hz sampling rate? All that data's gotta go somewhere). Ensure you've got enough space on your laptop or computer to store it all.

Audacity is helpful in that it can detect the free space on your computer and tell you how many hours of recording you've got room to store. It'll show you on the status bar at the bottom of Audacity while you're recording.

Saving a Project: Saving's as easy as going File -> Save Project -> Save Project As... Find a good location, give this project its own folder, and save.

Naming conventions: I'm going to assume you're creating a file as part of a larger project of an audiobook. If so, I recommend a regular naming convention. This makes it easy to find what you're looking for. Also, when you create your final .mp3 files for uploading to your distributor, they'll have a preferred naming convention.

I record scenes individually. They tend to be about 20-40 minutes in length. My own writing style has about 2-4 scenes per chapter. Thus, my naming convention for each scene will be, "booktitle_2-3.aup3" being Chapter 2, scene 3 of Book Title.

ACX, INaudio and others prefer that each sound file not be too short and not any longer than two hours. (Personally, I think a two-hour chapter is pushing it.) As a listener, I like an average size of 30 minutes or so. Anything longer than an hour feels awkward to me and is longer than the commute on which I'm listening to the audiobook.

From the production side, the longer the file, the larger it will be. Sometimes computers throw hissy fits if they have to edit too large a file. There's only so much space in your RAM.

If you have short scenes (like, ten minutes or less) consider having two or three of these short scenes in a single file. We can talk more about this in a post-production module.

Backing Up: Another thing you will want to do is export your raw .WAV file. This is an important backup. Should anything happen to your Audacity project file, you'll still have a copy of the raw .WAV you created. It's hard work laying down audio track. I'd hate to see you lose several hours' worth of performance work to a computer mishap.

I recommend exporting as a .WAV file. .WAV is a lossless format. This is as raw and uncompressed as you can get for a digital sound file. While they are large, they are also the truest, as no data is lost due to compression. (Thus, the "lossless" quality.) Only caveat is to ensure you've got enough storage space for these .WAV files. They can be quite large.

Also, .WAV is a universal format that will play on pretty much everything. As we're saving this just in case we need to go to backups, this is the best format to save in for sound editing.

To export your .WAV file in Audacity, go File -> Export -> Export as WAV. I recommend keeping the same naming convention as your Audacity projects, to make it easier to find, should you need them later.

I recommend exporting your WAV files to a location other than where your Audacity project is stored. And by all means, keep a copy on a different device other than your working laptop. Should something happen to your laptop, at least your data will be safe. (I speak from experience.)

I work in the computer industry as my Day Job. We're paranoid about backups. Keep a working copy on your laptop, and back up your work daily to another device, whether that's another computer, a server, or if you have the space, to a cloud. Even a thumb drive or a CD will do. Just make sure it's physically separate.

Once you have created and saved a sound file (and exported a WAV for backup), you're ready to edit.

2.3 Editing

Editing a soundtrack file is, essentially, removing all the unwanted bits and leaving the wanted bits. It's rare that you'll be adding anything to a file. (exception: you accidentally removed a wanted final plosive consonant sound, thinking it was unwanted noise; you want a bit more space between phrases, and you accidentally edited out too much.)

Later on in this book I refer to a YouTube video called "How I Record an Audiobook, from Go to Whoa", showing my entire audiobook recording process, including the editing steps I describe here. You can watch it now if you want, or you can read through this editing section first, then watch the video.

Youtube.com:
https://www.youtube.com/watch?v=sFy8cDZ_XxU

I have an unedited file of a reading from Pride & Prejudice. Feel free to download from https://heidikneale.com.au/AGRYOA/ (or see page 182 for link) and listen to it in all its raw glory. It's a good six and a half minutes long.

For playing with: **PP_raw_reading.wav**

For listening only: **PP_raw_reading.mp3**

I'm going to take this raw .WAV file, apply the following editing steps and create a lovely, tight file free of unwanted distractions to the listening experience.

First thing I do is check my track is in mono, not stereo. While I have the option of recording in mono (stereo is default setting for Audacity), sometimes I forget, and leave the settings in stereo when I record. You'd think I'd notice, but after years and years of working with multi-track projects for music, I sometimes forget that audiobooks are usually served up in mono, and I really should be recording my tracks in mono. At least if I do forget and end up with stereo, It is easy enough to mix down. (Tracks -> Mix -> Mix Stereo down to Mono)

When I edit my files, I take three passes:

1. I listen through the file. This familiarises myself with it, so I know how many repeat mistakes I've made, or any undesirable noises, etc. I don't actually touch anything at this point. I'm simply reviewing for quality. If I come across performance issues, I'll either redo it with a punch-in for a single issue, or I'll re-record the whole thing if I don't like any of it, or it's too much work to punch-in multiple times.

If I have time, I'll do this step immediately after I record this project, just in case anything has to be redone. This is a quality control issue. Same room, same space, same air pressure, same voice quality. Fortunately, it's rare that I need to re-record a track now, but in the early days, I've had a few chapters I re-recorded in their entirety.

Your mileage may vary, especially if you haven't established a standard operating procedure. Once your body gets into the flow of narration, you may find you make fewer mistakes.

Once I have a track I'm happy with, I can proceed in my editing.

2. I go backwards through the file from the end to the beginning and remove the obvious unwanted noise. There's two ways I do this:

Removal. Removal is simply highlighting the noise I don't want and pressing Delete. This removes that bit and closes up the sound file, timewise:

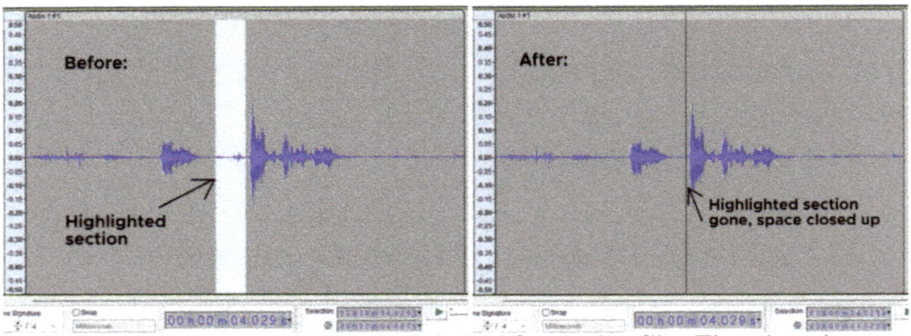

Figure 11: a breath mark deleted from the file. The time is deleted as well, and the dialogue is moved up time-wise.

Replacement. Because timing is everything, sometimes you don't want to truncate pacing. Instead of deleting the unwanted noise, I replace it with room tone.

First, I'll copy about 0.25 to 0.50 seconds of ambient sound floor (room tone) from the end of my file. The amount of room tone I select depends on the pace of my narration. If it's slow and languorous, I'll choose a longer selection (like a half-second). If I'm editing a faster-paced section, I'll choose a quarter second. This is because I want to preserve the pace, with the natural pauses that come between the wanted dialogue.

Next, I select the noise I want to remove. It doesn't matter how big or small this selection is. It does not have to match the time length of my selected room tone.

Finally, I will copy/paste my room ton over the unwanted noise. However big my selection was, it is now replaced with about 0.25 seconds of room tone.

Audacity holds your last copy selection in its clipboard, so I can simply paste-paste-paste as needed. This helps preserve the timing of the sound

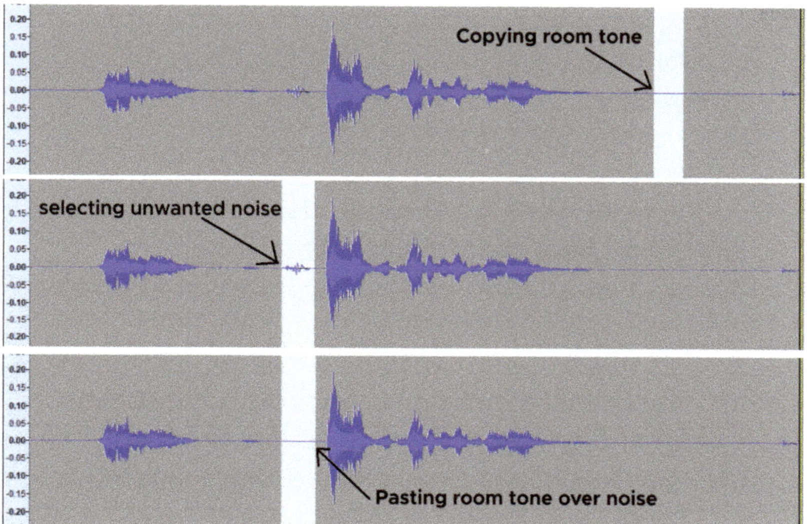

Figure 12: copying and pasting room tone over unwanted noise. Unwanted noise now gone.

In this editing pass, I do this by sight. After hours and hours of watching sound files, I got to recognise what a breath looks like, what an unwanted pop and other noises look like.

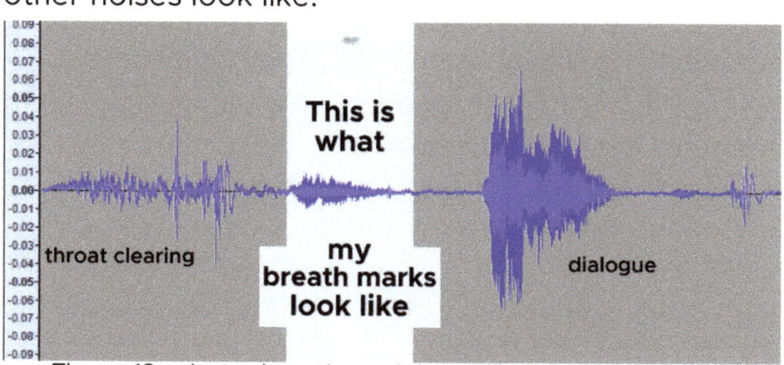

Figure 13: what a breath mark looks like.

I clean out the obvious ones, plus tighten up any extra-long gaps of silence.

If I see something and I'm not sure what it is, I'll listen to it to confirm before removing it (Someday I may employ an AI analytical tool to catch most of the undesirables, mark them or replace them with room tone, but until then, I do it by hand.) If it's unwanted, out it goes. If not, it stays.

3. I listen forward through the file and clean up anything I hear that I don't want. Unwanted dialogue (as you will notice in my sample file), repeat removals, coughs, drinking noises, any ambient sounds popping up, etc that I didn't recognise on my first pass or clean up on my second. I also listen for pacing and timing. For example, Mrs Bennet's dialogue has a fast pace, whereas her husband is a bit slower, with longer pauses between his phrasing.

One thing I don't clean up at this stage is the length of the room tone at the beginning and ending of the file. I'll tidy up any other noises, but otherwise leave the room tone alone. This I leave for post-production.

Once in a rare while, I'll need to apply a certain effect to a small selection of sound. Usually, this is fade-in or fade-out, as I take the edge off an extra-sharp attack or release. Sometimes it's the manual removal of a single pop or click that crept into the recording. It's best to fix these little things as I find them for the sake of time efficiency.

After I do that cleanup pass, I'll listen to the file one more time. If I hear any further issues (rare), I'll clean them up. Once I get a soundtrack I'm happy with, then I can go on to post-production and master my file, ready for upload.

Have a listen to the edited sample of Pride & Prejudice. (see page 182 for link, or visit https://heidikneale.com.au/AGRYOA/)

Download sound file: PP_edited_reading.mp3

I've cleaned this up. It's neater, tidier and much nicer.

One noticeable difference is that it's now only 4:51 minutes long. It's a good 75% of the raw length, my having cut out a minute and a half of unwanted sound and silence. Now that we've got all the needless breaths and mistakes out, we're ready for post-production.

If you want to have a go at editing my Pride & Prejudice raw file, by all means, do so. See the link at the beginning of this section to download a raw .WAV file of my Pride & Prejudice excerpt.

Homework

Take one of the sound files you recorded before, or record a new one. You'll only need a few paragraphs of something.

Listen through it. Do you hear any noisy breaths? Any coughs, or other unwanted noises? Normally our brains filter those out of our everyday lives, but on a recording they stick out like the butt on a wombat.

Once you're aware of these subtle little noises, you'll be able to spot them in your recordings and remove them.

Stereo vs mono

When it comes to fully immersive sound, stereophonic sound is the beez kneez. I love stereo! I've had so much fun with it in the past.

Stereo sound takes advantage of the fact humans have two ears. This dual sound-capturing system we've got on the sides of our heads (aka binaural reception) enables us not only to hear sound but triangulate its location. Stereo can give a marvellous sense of depth to a piece of music or a movie soundtrack. Surround Sound in theatres (like mine) gives you the sense of being immersed in the movie. It lends itself very well to escapism.

You'd think that stereo, being All That, would be the world standard for all sound.

But for audio books, the preferred format is actually....

Mono.

Wait. Mono? Didn't mono die out in the 60's with 45 records and flower power? Well, for music, yeah. But for narration, mono has some advantages.

- A sense of intimacy. When you're listening to a story, it's just you and the reader. The experience of an audiobook should feel like the narrator is sitting right across from you, reading to you. While stereo gives sound a sense of space, mono gives sound a sense of immediacy. This makes it close and personal.
- File size. Stereo is, essentially, two sound tracks. Mix these down to mono, and you effectively halve your file size. As we're working with moderately large files, a smaller size ratio per play time is beneficial during post-production. Some of the effects I apply in post has a size limit of what they can handle. 40 minute mono track? No problem. 40 minute stereo track? Uh, problems.
- Cleaner sound. Stereo is designed to give sound a sense of space. This means the left and right channels could vary in comparative signal strength. This can sometimes affect voice clarity. It can also introduce a higher noise floor.

- Technical reasons. As a mastering process, stereo should be a final step. But for audio books, once you upload your files to your distributor, it may not necessarily be the final step. Many companies, such as ACX, will take your files and convert them into other formats for their different markets. Stereo files have greater potential for issues. ACX prefers mono and strongly suggest we upload all our files as mono.

If you can record your files in mono, do so. The Audacity device toolbar has the option for selecting stereo or mono for your microphone.

And if you're like me and forget you're recording in stereo, you can still mix down your track to a mono track through Tracks -> Mix -> Mix Stereo down to Mono.

Fun with Stereo

Here's a short piece I composed a long time ago where I played with stereo effects. This piece is best listened to with headphones. Otherwise, if accessing this on your mobile phone, put the phone under your chin for best effect.

Listen to sound file: WHC Ambience2.mp3
(see page 182 for link, or visit https://heidikneale.com.au/AGRYOA/)

Why you shouldn't be the one to edit your audiobook:

- technical stuff is too scary
- editing process is too long, too boring, too isolated

My condolences. However, I hope you had a chance to play with the files I provided for download, or created files of your own. Editing as an exercise is rather simple. Don't let my long explanation daunt you from doing some simple editing. In fact, it is so simple, some people find it tedious and dull.

But if you've given it a go, and the thought of picking out all the unwanted bits of drift from the soundtrack is too much for you, that's okay. You can outsource editing and other post-production steps to someone else. Plenty of sound engineers online who are happy to take your raw vocal track and clean it up for you. Yes, this will cost money, but not as much as it would have cost to pay a narrator as well.

Chapter 3 - Narration

Welcome to Narration

This chapter is all about what you bring to an audiobook–your voice. If you, the author, plan on narrating your own book, this chapter is for you.

I'll talk about how narration works, especially what to expect from yourself if you're narrating your own book. I've got some sections demonstrating speaking speed plus a few other goodies I hope you enjoy.

Finally, I'll get into why to edit a sound file, what to look for, what to leave in, what to take out, and what's so beyond hopeless, it's better to simply re-record the track and try again.

3.1 Narration Overview

What is Narration?

Narration is a spoken performance. Acting skills greatly enhance this performance and can make the difference between an okay and a great delivery. Speed, vocal quality, and room quality all play parts in good narration. I have a YouTube video "Audiobook Narration Voice Techniques for Authors" in which I give a few tips for lifting the quality of your voice, including voice positioning and reading speed.

Youtube.com:
https://www.youtube.com/watch?v=c_DSH9vwclM

Plenty of other experienced narrators also offer videos that can help lift your narration game. I include some links in the back of the book.

Not an actor? Don't worry. Listeners will forgive much if a book is "read by the author". With a bit of practice and feedback, you can train your voice to suit personal narration.

The absolute best way to improve your reading performance is practice. Practice, practice, practice! Read out loud. Read lots. Read lengthy. Can you do an hour stretch? The more out-loud reading you do, the more skilful you become. A good way to practice is to read others' books out loud. Bring back storytime!

If you want to narrate your own novel, get used to reading aloud. Record yourself reading at least an hour. Go back to listen. What do you notice about your voice at the ten minute mark? The thirty? The full hour? Does your tone change? How about your pitch and articulation? Does your speed change? Have you grown tired and it shows? Narrating can be hard work. Any marked differences within that thirty minutes is something to note and consider.

Some people new to narrating might find that they can only narrate for about fifteen minutes before their quality suffers. That's fine. If you can only do fifteen minutes, do fifteen minutes. Don't push yourself to try and fit everything into a single hour session. You can break it up into fifteen-minute sessions, with rest time in between. Quality is more important than time efficiency.

Your style is your own. Own it! Like your writing, you have a voice that belongs to you. As long as you're aware of technical issues such as speed and clear articulation, good chances are your voice is good enough to narrate your own books.

My first narration project was a 30K novella. I learned a lot through that experience, including learning when I need to re-record a scene. While I finished that project, I trunked it. I suspect I'll want to re-record the whole thing before I release it in the wild. My second and third projects were much better. Only then did I dare tackle a full-length novel with my fourth project. By then, I'd established my methodology and now I generate consistent product with fewer errors.

Give yourself a bit of practice before you tackle your first Magnum Opus. It is much better to learn from your mistakes with a smaller project, than learn in the middle of a novel, and suffer from quality inconsistency or loss of production time as you redo a project.

For the purpose of this book, I encourage you to get a short project done. For a first project, choose a short story, or a chapter from your book. Get used to sitting and recording, and bringing a finished product to fruition. You should be able to handle a project of about one finished hour, maximum.

Narration from You

What can you expect from yourself as a narrator? What makes you ideal, even if you're not a trained actor? What can you do to give a good performance and avoid making your listeners cringe?

Your Advantages

- You've got soul. Nobody knows this book better than you. You know it intimately. The characters are your babies, and nobody has a handle on them and the story and everything better than you.
- You have an emotional connection. You're naturally excited about the exciting scenes. You're passionate about the passion. You're empathetic during the sad scenes. You know what makes you cheer, what makes you afraid, what makes you sad. You bring this emotional connection to your reading. It will show.
- You pronounce all the names right.
- Nobody cares what accent you use. Use your natural accent. Don't think you must put on an accent to suit your novel. It's part of your brand as an author. Are you reading your Regency Romance with an Aussie accent? Okay! You get a puppy pass for accent simply because You Are The Author. Nobody's expecting a winning

performance. They're expecting the author to read them a rollickingly good story.

Things You Need to Get Right

If you're going to narrate your book, there's some things you need to get right to deliver a good performance.

- Warm up. It's best to give yourself a good ten minutes warmup narration before you actually record. This is because it can take some time to get into your natural rhythm. Once you're in this rhythm, you'll flow well, and your product will be consistent from recording to recording.
- Keep in mind that for a full-length novel, you'll be looking at possibly twenty or so hours of raw narration. You don't have to do it all at once; you probably can't. But as you break it up, keep quality issues in mind.
- Consistency of voice. Does your voice sound the same from hour to hour, from day to day? Inconsistency in voice can cause quality issues. You don't want to sound bright and bubbly one day and miserable the next. You need to be you, the authentic you, from chapter one to chapter last. Figure out how long you have a good performance in you. Record in that time space. Stop before you talk yourself hoarse.
- Slow Down. Your performance will improve if you can keep your reading speed around 150wpm. Give your words time to shine

How long will your project be?

Audiobooks, on the average, tend to be about 8000-10,000 words per finished hour. A full-length Romance novel might be about eight finished hours or so. The raw files that produce one finished hour of narration can be anywhere from 1.2 hours to 2 hours in length, depending.

3.2 Your Voice as a Tool

My voice is the primary tool I use for narrating an audiobook.

A good presentation voice is strong, easy to understand and pleasant to listen to. For most people, a good presentation voice is a skill one builds up. You can learn to wield it well. When you have some mastery over your voice, you can use it to add emphasis, give emotional resonance or create certain impressions.

What does your voice sound like to other people? How you think your voice sounds and how it actually sounds will be two different things! This is due to the resonance of your skull changing your perception of the voice, versus you listening to a recording of yourself. Everyone, upon hearing their voice recorded, often say, "Do I really sound like that?"

Yes, you do. And this is also how others hear you.

Homework

Go find that intro you recorded and posted for homework back in Chapter 1.1 (p32).

Listen to yourself. What do you notice about your voice?

- Do you like how you sound?
- Do you have a high-pitched voice or low-pitched?
- How's your speaking speed?
- Do you speak slowly or quickly?

- Any other qualities you notice?
- Is there anything you don't like about your voice?
- Do you sound too soft?
- Are you too nasally?
- Do you have vocal fry at the end of your sentences?
- Is there anything you wish you could change?
- What is the best thing about the sound of your voice?
- What qualities do you want to keep?

Lifting Your Game

Here are some techniques and qualities I consider when using my voice.

1. SLOW DOWN. One of the most common errors an inexperienced narrator can make is speaking too fast. Sometimes this is because of nerves. Other times it's because of trying to communicate too much information in too short a time. Also, your articulation may suffer under greater speeds as your oral muscles can't quite keep up with your brain. You don't want to muddy your words together. I like to slow down. I aim to make my words clearer and easier to understand.

Good articulation is preferred over a faster pace. On average, a slow speaking speed will be about 100 words per minute (wpm), a medium speed is about 150wpm (my preferred comfort range), and a fast speed is about 200wpm. When starting out, there's nothing wrong with aiming for a slower speed. Accuracy and speed will come naturally with experience and practice. But in the beginning, slower will be easier for your audience to understand, and you will make fewer mistakes. Fewer mistakes means less editing or re-recording you have to do.

See Chapter 1.3 for more information about speaking speeds and Words Per Minute.

2. Articulation. I want my consonants smooth and my vowels clear. I want my words to be nicely formed and not muddled up. By taking a bit more time to slow down and give good articulation to my words means I have a clearer, more open sound. After all, I don't want my words to slur together or my tongue to trip over them. This means I need to ensure consonants are solid and vowels are distinct.

Good articulation is a muscle thing. You've got several muscles governing how your mouth shapes sound, from the orbicularis oris muscles about your mouth, to your tongue, and even the diaphragm deep in your chest that controls breathing. Later on I can give you some fun, silly exercises designed to strengthen these muscles and give you greater control.

3. Voice positioning. My voice resonates at different locations within me, depending on where I'm focusing my energy. Certain placements are better for me, as there's a better energy/volume ratio. Where I place my voice within my head makes a big difference. I can then spend less energy and still get a strong-sounding voice.

Ever hear someone with a gravelly, throaty voice? How about a nasally whine? (And don't get me started on vocal fry!) Imagine listening to that voice for hours. Poor quality performance can detract from the beauty of your audiobook.

When I speak, I want to give my listeners a good experience for time we spend together. I want my voice to soothe, to carry them, to be pleasant to the ear. I don't want it to distract from the book itself. Positioning my voice so it sounds like it's resonating from the front of my mouth gives me the best sound. I'll show you a few techniques in a video so you can get the same positioning and sound good.

A benefit of good voice positioning is your voice will last longer. For me, good voice positioning means I can spend hours speechifying without going hoarse (laryngital fatigue). This enables me to spend an afternoon laying down hours of raw track.

Again, good muscular development will help with this.

4. Vocal fillers. This is known as speech disfluency. These are words like "Umm... Uh... well... Hmm..." that we use to fill in the spaces between our phrases. We all do them, sometimes to buy us some thought time, and sometimes to give a sense of continuation during conversation so listeners don't assume we're finished speaking.

When I'm narrating an audiobook, my words are sitting in front of me, I don't have to think about what I'm going to say next; I just read the lines. Smooth, pure, and without any extra sounds. I let the words shine.

Do not be afraid of a little silence. Silence serves you well.

5. Strange Mouth Noises. We're talking plosives, pops, clicks and more. Yep, these happen. Again, if you're not used to listening to your voice, you might not realise you're doing them.

A plosive is a consonant like P or B that gets pushed out with air. This can be problematic if you're speaking with a microphone. Mics are these lovely, sensitive devices that detect minute changes in air pressure. If you suddenly give it a strong puff of air with a P or B word, it will make you sound harsh.

Plosives are a bad habit most English speakers pick up. French speakers don't have this issue because they soften the impact of their plosives. I had to practice my plosives in French Immersion class by pronouncing P words into a tissue held up before my face. If the tissue moved, I was pronouncing it too hard.

Sibilance is the sound of S. Try words like: duchess, hiss, kiss, susurrus... you get the picture. S sounds can be extra-harsh if you push them too hard with breath. Again, this extra air can be problematic for a microphone. Sometimes sibilance is caused by drawing too deep a breath too quickly. The inward hiss can be intrusive.

Other strange mouth noises. The mouth is a wet, gross place. When it's less wet, it's even grosser, because it gets sticky. When that happens, "wet" mouth noises are made. They're sticky and sound like the snapping of chewing gum and are most unpleasant to hear. A plentiful bottle of water, less facial tension and awareness usually clears this up.

Pops and clicks can come from non-word sounds, or an extra harsh air push on certain consonants.

The good news is, you can train yourself out of most strange mouth noise habits. The first step is awareness that they are happening. I recommend recording yourself reading a five-ten minute passage of text. Listen to your recording and make a note of what kinds of sounds your mouth makes.

Personal sounds. Sneezing, sniffing, coughing, and clearing your throat. While you can't completely control when these things happen, you can reduce the chances of them happening. When they do happen during narration, pause, and let the event happen. Then once you're done sniffling/sneezing/coughing/shouting at the cat, pause for another moment, then resume your narration. All this extra stuff can be edited out later.

Before you sit down to narrate:

- Grab a tissue and blow your nose. This reduces your chances of needing to sniff during recording.
- keep hydrated. This will prevent wet mouth noises, coughing due to dry throat and a whole slew of other issues.
- avoid dairy products a few hours before narration. Many dairy products, especially fresh milk, encourages thicker mucus of the throat. This can interfere with clear sound and increase your chances of needing to cough or clear your throat.

6. Inflection. Nothing is more boring than a flat-pitched voice. I once attended a convention where the author Guest of Honour gave an hour-long reading from one of his novels. Man, did he drone on. More than one audience member drifted off. It was awful. Pitch variation and tone patterns can add interest and variety to your speech.

Here's some simple inflection techniques:

- At the end of a question, pitch should rise up.
- At the end of a statement, pitch can drop.
- A pitch rise can emphasise a word, indicate surprise or indicate lightness.
- A pitch drop can add gravity or disappointment.
- Don't forget to include pauses in your inflection. Beats and rhythm also play a vital part in a good performance.

In the next few sections we'll look more into these techniques and qualities, including some practical exercises to help you lift your game.

3.3 Your Speaking Speed

Most of the time, I will tell new narrators to slow down. Just going to tell you that right now. A lot of people tend to speak too fast when performing or speechifying. Usually, this is nerves. Other times it may be excitement over a topic, or maybe a desire to get through the speech as quickly as possible. Talk too fast, and you will do no one any favours, least of all your audience.

What's a good speechifying speed? About 140-160 words per minute (wpm).

This section includes links to audio examples

Homework:

How to Find Out Your Speaking Speed

Here is a one-hundred-word section taken from a short piece called "The Rainbow Passage"[1]. The Rainbow Passage is an often-used public-domain passage from the Voice and Articulation Drillbook used for English speech evaluation. It is used in speech therapy, voice training and more to help improve diction and articulation. The whole passage utilizes all 40 phonemes of American English in a way that reflects their representation in everyday conversation. I include the full text of the Rainbow Passage at the end of this section. Feel free to use it as a warmup or for your own voice training exercises.

---

[1] *Fairbanks, G. (1960). Voice and Articulation Drillbook, 2nd ed. New York: Harper & Row. pp124-139.*

Record yourself reading the following 100-word extract from The Rainbow Passage aloud and time yourself with a stopwatch.

"When the sunlight strikes raindrops in the air, they act as a prism and form a rainbow. The rainbow is a division of white light into many beautiful colors. These take the shape of a long round arch, with its path high above, and its two ends apparently beyond the horizon. (51 words)

There is, according to legend, a boiling pot of gold at one end. People look, but no one ever finds it. When a man looks for something beyond his reach, his friends say he is looking for the pot of gold at the end of the rainbow." (99 words)

How long did it take you? Compare your speed with this table:

Time	Speed
60 seconds	100 wpm
55 seconds	115 wpm
50 seconds	133 wpm
45 seconds	150 wpm
40 seconds	165 wpm
35 seconds	183 wpm
30 seconds	200 wpm

Table 1: length of time to read the Rainbow Passage excerpt determines speaking speed.

If you read it any faster than 35 seconds, try again and slow down. You want to aim for about 40-50 seconds.

Like I said before, between 140 and 160 wpm is a nice, comfortable speed that most listeners will enjoy. You won't sound too slow, and listeners will have time to absorb your ideas.

Yes, I know that you might think you're speaking too slow while reading The Rainbow Passage. (You're not, I promise.) This is because you're most likely used to a conversational speed around 180 wpm, maybe faster.

Have a listen to me at 140 wpm and 160 wpm. Which speed do you prefer?

Listen to sound files:

The Rainbow Passage at 140 wpm:

Rainbowpassage_140.mp3

The Rainbow Passage at 160 wpm:

Rainbowpassage_160.mp3

(see page 182 for link, or visit https://heidikneale.com.au/AGRYOA/)

One way of reducing your speed without dragging is to separate your phrases a bit more. You don't want space between words, but between phrases.

That said, there is such a thing as too slow.

For example:

Listen to sound file: tooslowspeech.mp3 (see page 182 for link, or visit https://heidikneale.com.au/AGRYOA/)

"This is an example of a speech slowed way too slow. There is too much space between words for the human brain to make logical sense of them. The ideas lose their meaning and you will forget what I have just said." (42 words) 40 seconds, or ~60 wpm

"This is an example of a speech given enough space. Instead of spreading out words individually, phrases are kept within liminal space to express an idea. Space between the phrases allows the human brain to parse the idea. Not a bad start." (42 words)

33 seconds or approximately ~70 wpm. This is half of the recommended bottom limit of 140wpm for speechifying.

Even though that second passage is a bit faster, it is bordering on too slow to make much sense. It would be much more comfortable to be able to speed it up, even to 100 wpm.

Now that you've heard me read a really slow passage, go back and listen to your recorded Rainbow Passage bit. Doesn't sound too bad now.

The human brain has an innate sense of time. There's a point where the brain will make a connection between two beats (or syllables or words) that is chronologically linked. This is called Perceptual Presence. (Think of Perceptual Presence as a fancy way of saying, "being in the now" or "being in the moment.")

But that space between those two beats can only be so big. If that space gets too big, then your brain doesn't link the two beats temporally. It will separate them and turn them into two distinctive objects, instead of linking them as a joined pattern pair.

Part of you delivering a talk is to communicate complex ideas. You'll drop one idea in someone's head, then follow up with a second idea that will deepen the first thing they heard. You want their brain to link the two ideas together as a matching pair (or trio).

While I usually am telling people to slow down, there is such a thing as too slow.

Want to go hardcore? Here is a really awesome video of New York bass player Adam Neely explaining the bottom end of slow. While this is mostly about music, the rhythmic structure will also apply to spoken speech.

Visit YouTube: https://youtu.be/afhSDK5DJqA

You can listen to the whole thing (though the last ten minutes might not interest you at all). It is a good fifteen minutes long, but some points that might interest you are:

- Subjective Rhythmisation
- Perceptual Presence (Can abandon at 3:40 if you wish)
- Liminality (Can abandon at 5:20)

If slow is too slow, then what is too fast?

That depends on a few factors like clarity and complexity of ideas. Some trained narrators can ramble on at 200 wpm and still be sharp. Auctioneers rip along nicely at a good 400wpm, sometimes faster!

The main reason for not going too fast is that most people, when they speed up, lose diction control. When that happens, they don't sound clear. Their orbicularis oris muscles aren't used to going too fast with precision. They start to stumble, to fumble, and your sound gets muddy and garbled. While you can train up your mouth muscles to be able to handle clear speech at a quick pace, until you master this, it's okay for you to keep to a slower pace. Clarity in speech is your friend.

Find your normal speaking speed. Your narration speed will be slower than that, but not so slow that you can't keep your phrases within liminal space.

In a later section I'll present some speech and diction exercises to help you tune up and master that lovely tool you've got–your voice. Be prepared for a bit of silliness.

The Rainbow Passage (complete)

When the sunlight strikes raindrops in the air, they act as a prism and form a rainbow. The rainbow is a division of white light into many beautiful colors. These take the shape of a long round arch, with its path high above, and its two ends apparently beyond the horizon. There is, according to legend, a boiling pot of gold at one end. People look, but no one ever finds it. When a man looks for something beyond his reach, his friends say he is looking for the pot of gold at the end of the rainbow. Throughout the centuries people have explained the rainbow in various ways. Some have accepted it as a miracle without physical explanation. To the Hebrews it was a token that there would be no more universal floods. The Greeks used to imagine that it was a sign from the gods to foretell war or heavy rain. The Norsemen considered the rainbow as a bridge over which the gods passed from earth to their home in the sky.

Others have tried to explain the phenomenon physically. Aristotle thought that the rainbow was caused by reflection of the sun's rays by the rain. Since then physicists have found that it is not reflection, but refraction by the raindrops which causes the rainbows. Many complicated ideas about the rainbow have been formed. The difference in the rainbow depends considerably upon the size of the drops, and the width of the colored band increases as the size of the drops increases. The actual primary rainbow observed is said to be the effect of super imposition of a number of bows. If the red of the second bow falls upon the green of the first, the result is to give a bow with an abnormally wide yellow band, since red and green light when mixed form yellow. This is a very common type of bow, one showing mainly red and yellow, with little or no green or blue.

Fairbanks, G. (1960). Voice and Articulation Drillbook, 2nd ed. New York: Harper & Row. pp124-139.

3.4. Audiobook Narration Skills

Narration is a type of acting. Professional narrators are a form of voice actor. While you don't need to be a trained actor to be able to narrate your own audiobook, having some knowledge of how to better use your voice will improve your product. The most important thing is to be your natural self. Relax and let your personality shine through. You don't want to be stiff or formal or frightened. Let your words flow naturally, as if having a conversation.

Because it's impossible to fully describe good narration skills purely on the page of a book, I turned these tips into a YouTube video: Audiobook Narration Voice Techniques for Authors
Youtube.com:
https://www.youtube.com/watch?v=c_DSH9vwclM

Breathing

Proper breathing is important. When you breathe in you want to draw down with your diaphragm. When you do it properly your lungs will expand making your ribcage expand out. You'll also know you're doing it properly when your stomach comes out as well. You don't want your shoulders to lift because that's not really extending with your ribcage.

When you speak, you don't want to push too much air too hard. This can lead to breathiness, sibilance and strong plosives. You want a nice even push. If you find it difficult to maintain an even push you may need to strengthen your diaphragm.

One fun exercise singers use to strengthen their diaphragm is called "buzzy lips" or "motorboat".

Buzz your lips as if you're a motorboat or an exasperated horse. (Try not to laugh. Okay, you will most likely laugh. Stretch your face and try again.)

This one works by using the oris ocularis muscle around the mouth to resist against the pressure from the diaphragm. This resistance strengthens your diaphragm.

Voice Positioning

Too many people speak in the back of their throat or up in their nasal cavity. When you're talking in the back of your throat, it sounds a bit gravelly and low. This can cause voice strain. You don't want that especially if you're talking for hours and hours.

If you're speaking up in your nose, you sound a bit nasally and whiny. You really don't want to sound like Fran Fine from The Nanny.

The best place to position your voice is in the front of your mouth. How do you know it's placed properly? By humming. When you hum, your voice naturally places itself in the front of your head. You should feel it vibrate about your nose. (If you can't feel it, put your fingers on the bridge of your nose.)

Now open your mouth and move from a hum to a Maaah sound, keeping the vibrations in that same position.

Articulation

The tongue and the muscles around the mouth can get a bit lazy sometimes. When this happens articulation can suffer especially at high speeds. Tongue twisters are an excellent way to improve the responsiveness of your mouth in moving between different consonants. One of the favourites of actors is, "The teeth, the lips, the tip of the tongue."

This one is great because it positions your voice in the front of your mouth. Start off slowly, and pay attention to how your mouth feels around the words. Repeat this one over and over.

Little by little, speed up. If you find your consonants and vowels getting muddy, ease back on the speed. Mastery first; speed comes naturally with mastery.

Inflection

Do you suffer from flat tone? Do you drone in a monotone when you read? Do your listeners fall asleep from the sound of your voice?

Being mindful of inflection will solve that. Your pitch will go up when you ask a question. With a statement, your pitch will drop. You can emphasize a word in a sentence by raising your pitch in the middle.

I hope some of these techniques help you to improve as a narrator. For more exercises get on YouTube and have a look around. Plenty of actors and recording artists have posted some useful exercises that can help you become better at narration.

Meanwhile, record your voice and have a good critical listen. Pick a thing or two that you like about your voice, and make a note. Then choose something you'd like to improve about your voice and work on that.

3.5 Narration Session Tips

Ready to narrate? Go right ahead! Pick a scene in your story, get comfy in front of your mic, and hit record.

Here's six tips for having a great narration session.

1. Be at ease. When you narrate you want your body to feel comfortable and relaxed. When you are relaxed this will show in your performance. Some people like standing and narrating. If they are able to stand for an hour, more power to them. Many people find it easier to breathe, to move, to articulate correctly. But me, I prefer to relax in a semi-reclined position like a recliner. I'm afraid my osteoarthritis won't let me stand for a long time. By reclining, I can focus on my performance and not get all stiff and tight.

If you're not used to standing for long periods of time, sometimes your muscles will tighten up, starting with your legs and your core muscles and into your throat and everywhere else. You'll find your voice gets tight as well. You don't want that.

2. Wear comfy clothes. You want to be comfortable. You don't want tight clothes and you don't want noisy clothes. Yes, clothes can be noisy. Some fabrics, like crisp cotton shirts, or slithery fabrics like satin or nylon, can be noisy.

If you wear something soft like a knit then not only will you be comfortable but you'll also make less noise. You don't want extra ambient noise around you to ruin your performance. My favourite recording outfit is a soft cotton T-shirt and tracky-daks (sweat pants).

3. Warm up your voice. You don't want to go into a reading with your voice cold or stiff. Your voice uses muscles. These muscles will perform best when warmed up.

Warm up your voice for a few minutes with the text you're about to record. Not only will this warm up your muscles, but get you in the right frame of mind. I recommend reading the first three to five paragraphs. This will limber up your voice and give you consistency in your sound.

Warming up will also work out all the mucus and the tightness in your throat.

3a. Keep a bottle of water close by. Sip from it regularly. As you use your voice, it will dry out. By keeping your whistle wet, you'll avoid "wet mouth noises" and prevent strain.

4. Be familiar with your text. Yes, I know this is your book. However, your book might be very, very long or it might have been a while since you wrote it. While you know it overall, you might have forgotten the little moments. Before you narrate, sit down and look through the scene. You don't have to read the whole thing through, but glance through it and remind yourself: What is happening in this scene? Who are the characters involved? What part of their journey are they at? Are they happy, are they sad? Is a change about to happen? What is going on?

This will get your mind and heart into what you are about to read.

5. Slow down. When you narrate, remember to slow down. Give your brain and your voice space to let the words and emotions shine through.

6. You are not reading a book, you are telling a story. When people read stuff they'll go through and look at the text and drop into this monotone voice as they try to get through the text. This risks your voice becoming dull and tedious. This is the last thing you want. You risk lulling your readers to sleep. Unless you plan on becoming an accidental ASMRtist, you want to Tell a Story. You want to be lively and exciting. This is a good book! You wrote it. The things that happen to the characters should be interesting. You want to communicate that interest with the readers.

So don't "read the book", tell the story. Your story is full of emotional resonance. Bring that out when you narrate.

Homework

Select a few passages from one of your books, preferably ones with different moods, vibes or action sequences. You could do an entire scene or just a few paragraphs. Record yourself and listen to your performance.

It is a good idea to listen and review your performances as a narrator. Listen with a critiquing ear. Are you speaking slow enough, and not rushing your words? Do you sound clear? Are you bringing out the emotional resonance, or does your narration sound flat?

If you don't like what you hear, analyse this passage. Make a list of what you don't like, then address how you can fix it. If you are feeling brave, feel free to share a raw recording with a fellow author or three and ask for feedback. (I don't recommend sharing with friends and family, as they are biased and may not give you accurate feedback.)

Sometimes a few friendly folk on Audacity's Audiobook Forum (https://forum.audacityteam.org/viewforum.php?f=64&sid=f9af0b4f0795c7b2f5ff4d12b7740ca8) are willing to give some feedback, but they tend to focus more on production quality (noise floor too high, bass rumble, etc) rather than performance quality.

Do an editing pass on these sections as per Chapter 2 and save them. These will be excellent practice for Chapter 4: mastering your audiobook files.

Additional Resources

As a narrating author, here's a few handy resources by professional actors and audiobook narrators on performance improvement, including several free YouTube classes. I'll also include some fun stuff later in this chapter that can help you improve your performance and your vocal resilience.

- Patrick Fraley
  https://patfraley.com/pf/free/
  (I love his free lessons!)
- Narrators Roadmap
  http://www.narratorsroadmap.com/
- Tanya Eby
  http://www.tanyaeby.com/blog/2018/12/16/so-you-want-to-be-a-narrator
- Sean Pratt's test to see if you suit narration
  https://www.youtube.com/watch?v=NPzPi-_0Xi8

Chapter 4 - Mastering Your Audiobook Files

Here I start to get into the final steps of audiobook production. I've covered recording sound files and editing in chapter 2. Now we look at post-production which includes mastering my files to meet industry standards and saving them in the best format for uploading to my distributor.

This section will be very hands on. I'm going to give you lots of homework.

I want to see you try out my process so you can get a feel for what I do. You can play around with your mic and look for solution for creating a really good raw sound file. Best practice dictates that you create the best raw file you can and limit editing to a minimum. While you can make a good file sound better, you can't make a bad file sound good.

If you're thinking about future projects, now would also be a good time to look into Audacity's Audiobook Forum (https://forum.audacityteam.org/viewforum.php?f=64&sid=f9af0b4f0795c7b2f5ff4d12b7740ca8). It's a whole bucket of useful information and gives you access to some industry experts. Keep an eye out for a guy named Kozikowski. He really knows what he's talking about.

4.1 Mastering your audiobook files

Mastering a sound file is the filtering, prepping and tweaking required to refine the file and make it sound really professional.

Homework

While you're still experimenting with your setup, go ahead with the mic you've got and lay down some audio track, if you haven't already. Give it an edit as per Chapter 2.3 to remove unwanted noise, coughs, breaths, et al. Get a track that, for the limits of your hardware, sounds like the text from which you read it.

For this section, I want you to master that track to see if you can make it sound even better.

Why Master a Soundtrack?

Mastering is where I ensure my audiobook soundtracks meet the industry's stringent Quality Assurance (QA) guidelines. I need to improve the track without compromising its integrity and therefore improving its sound quality. My aggregator's QA department will knock me back if I sound like I'm talking out of a tin can.

Unlike editing, which removes bits and pieces of the soundtrack but otherwise leaves its integrity intact, mastering messes with everything. This is why it's so important to start with the best raw soundtrack you can. You can't fix a bad soundtrack, but you can make a good soundtrack better.

It's also good to keep processing to a minimum. Often using a filter or effect to fix one thing can mess up something else. Prudence will serve you well.

Audacity has a whole lot of Effects with which you can manipulate a sound track. From Fade In and Fade out, to pitch and time bends, filters, passes, echoes and choruses and more, there's a whole lot of nifty things you can apply to a sound file.

Doesn't mean you're going to be using the majority of them when you master, but it is fun to play with them just to see what they can do.

Check out the Effects menu under Audacity. Amplify and echo and phaser and stretch, fade in/out, pitch bend, time stretch, and do all sorts of fun things to that file, just to see what the tools do. You can really put a sound file through a wringer with all the tools available to you in Audacity. Have lots of fun with them now, because once you start mastering a sound file for the purpose of producing a quality-level audiobook, you'll probably not touch any of them. Voice production isn't about fancy effects, but ensuring a simple, pure sound.

Our purpose in mastering our sound files is to take the rough edges off, drop off that noise floor and make everything match without making it look like we've done anything. The steps in mastering are far more subtle than the fun Effects you've just played with. If we've done our work right, our efforts will appear invisible and you'll only notice the narrative performance.

Mastering a File

There are a few housekeeping things I do before I master my files.

First Step: I back up my edited .aup3 files. This is just in case I mess up my mastering. I do not want to make the gods of audiobooks cry by failing to create backups.

Second Step: I take all the scenes of my chapter and combine them together in a single Audacity Project, one chapter per project. I do this with a simple cut and paste.

Why I do this: to make editing easier, I tend to record scenes as their own separate projects. It just makes them easier to handle and edit when they're smaller. But once I'm done with editing, I combine all the scenes into a single project for each chapter before I master it. This is also to ensure consistent quality. If I mastered each scene of a chapter separately, then combined them together, that increases the chances of bad quality artifacts.

Let's master!

These are the steps I take to master a file. **I must do these steps in this order**, or I risk messing up the effects of one or other of my steps.

0. Rumble Removal.

Rumble is a low-pitch rumbling noise that you don't really hear (think unwanted subwoofer) but can affect your sound floor and the quality of your sound file. Removing it can help drop your sound floor and provide some clarity to your file.

Actually, I don't do this step often because I generally don't have rumble problems. But I know several other sound engineers who do. Because it's a prevalent enough step in the industry, I'm including it here for those who may need rumble removal.

If you're using your Zoom mic for a throwaway test project, you might not need to do this either, as rumble is lower bass frequencies (often at the bottom or beyond the range of human hearing) and your Zoom mic sounds like it's not terribly bass-heavy. But due to some fun, complex physics rules that sound waves like to follow, rumble can affect the quality of a sound file.

It's worth experimenting to see if rumble removal makes a difference for you or not. Create a sound file and make a copy. For Copy A, apply this Rumble Removal step, and carry on. For Copy B, omit Rumble Removal, and carry on with the rest of the mastering steps. At the end, analyse both files, and/or give them a listen. Can you tell the difference? Did removing rumble improve your quality? If it doesn't make a difference, don't bother with this step.

Select your whole sound file before applying the Rumble Removal step.

The easiest way to get rid of low-frequency rumble is to use the *Low Rolloff for Speech* setting in the Equalization tool.

In Audacity, go...

Effect -> Equalization. Select curve: *Low Rolloff for Speech*. You'll notice the filter change in the window.

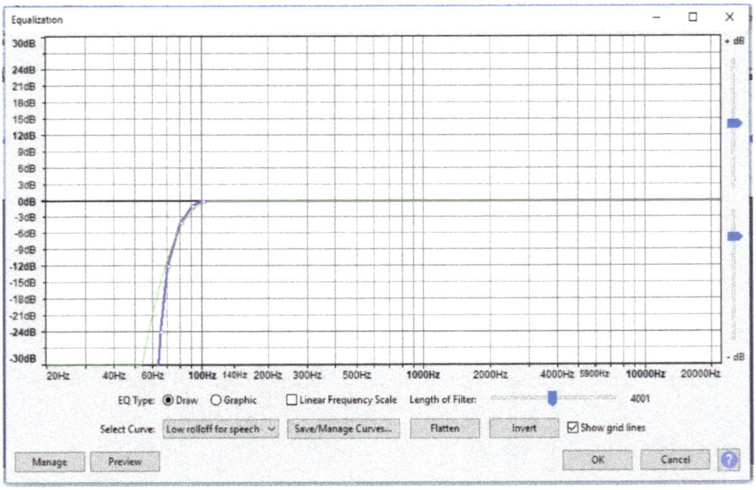

Figure 14: the Equalisation window with a Low Rolloff for Speech filter applied.

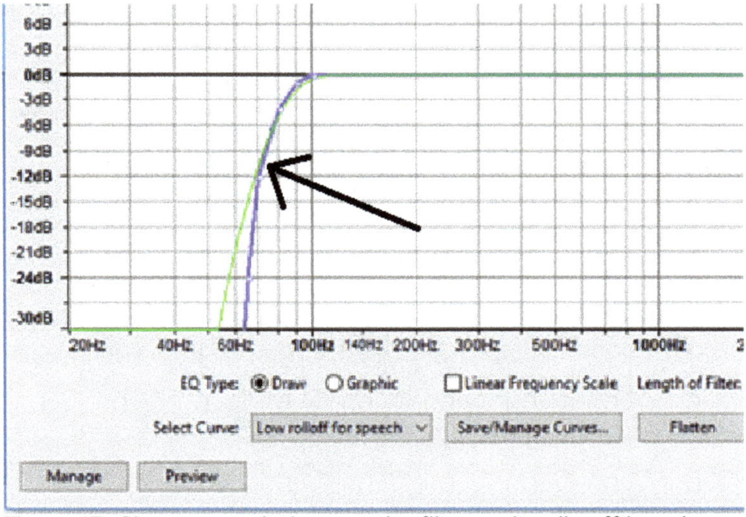

Figure 15: Close-up, pointing out the filter as it rolls off in volume, at about 50Hz.

Essentially, you're creating a high-pass filter that will let the higher frequencies pass through and will block the lower, unwanted frequencies. Human hearing can sense pitches as low as 20Hz, but generally, we only notice it about 40Hz or thereabouts.

Click OK to apply the filter to your selected sound track.

How do you tell if you need to remove rumble?

- You hear rumble in your raw track. It may sound like a low growl, earthquake, or a distant train passing.
- Your sound floor appears higher than -60dB (more later on how to tell what your sound floor is. You can use the ACX Check plug-in to find out. See Step 4 Quality Assurance)
- You notice a difference after you apply your rumble filter. Remember, you can undo a step immediately after you apply it. If you apply a rumble filter but

don't notice any change in the sound quality or the noise floor, undo it.

1. Noise Reduction

This is the first and most important step in mastering a sound file.

This gets rid of all the ever-present background noisiness. It also drops your noise floor. Noise Reduction is a standard tool that comes with Audacity. Our goal is not to completely silence the space between our narration, only to drop the noise floor to acceptable levels. Dead silence sounds unnerving. Humans are used to ambient noise about them. We want some ambient sound, but we don't want distracting noise like line hiss, etc.

In Audacity, go...

Effects -> Noise Reduction...

This step has two parts.

a) First, sample the noise I want to reduce, in this case, the leading room tone and ONLY that.

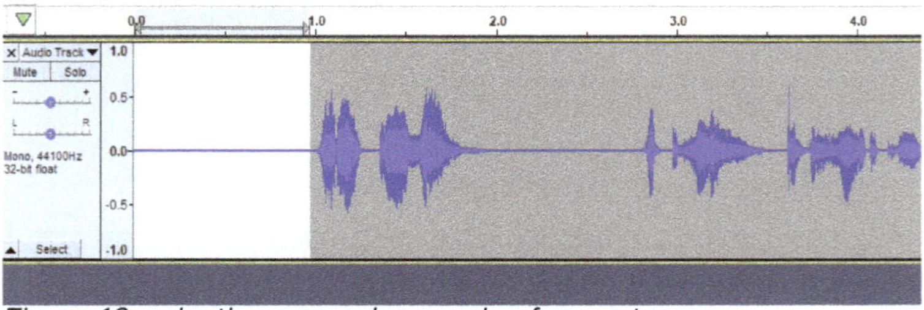

Figure 16: selecting several seconds of room tone.

I select several seconds of clean room tone from either the beginning or the end of the file. While that's selected...

In Audacity I go to *Effect -> Noise Reduction...* A window pops up. Notice the Get Noise Profile button is highlighted?

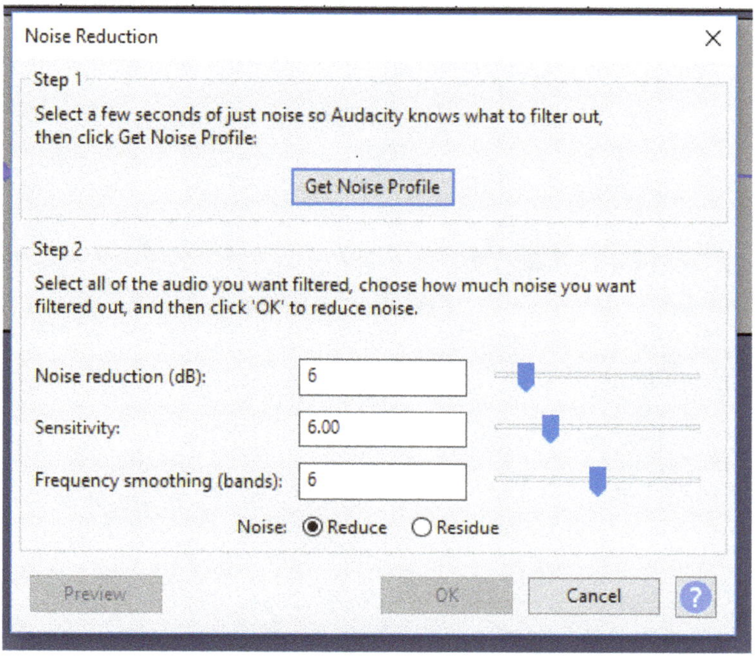

Figure 17: The Noise Reduction panel, showing the two steps needed to reduce noise.

I click “Get Noise Profile”. This samples the noise from the selection I made earlier. Now we’re back to the soundtrack.

b) Reduce the noise.

Now I select the whole soundtrack. (ctrl-A is a good hotkey combo to do this.) While that’s selected...

I go to *Effect -> Noise Reduction...* Time for Step 2.

I configure the noise reduction criteria I want. Audacity will remember these for next time. I don’t go for the default. Instead, I choose settings fondly known on the Audacity Forum as the "Noise Reduction of the Beast".

Essentially:

- Noise reduction: 6
- Sensitivity: 6.00
- Frequency Smoothing: 6

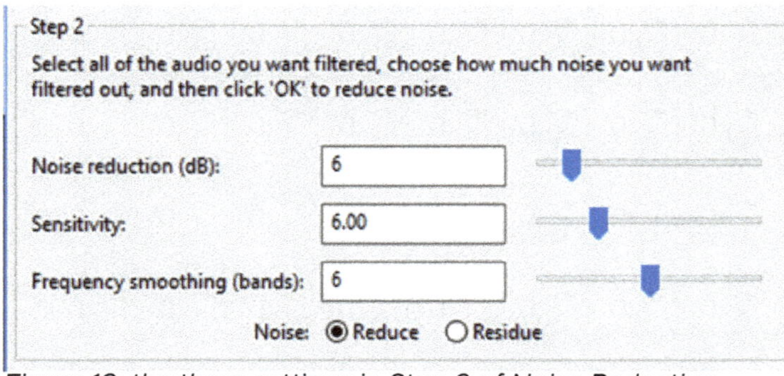

Figure 18: the three settings in Step 2 of Noise Reduction.

Depending on your situation, you can play with these, maybe going 6,6,8? Find what works for you and stick with it.

If you want to see how this level will work, you can click the Preview button and have a listen. Otherwise, click OK and enjoy your new, quieter, soundtrack.

2. RMS Normalize

RMS stands for root-mean-square, or the average loudness of a sound file. The ACX Quality Assurance standards (which everyone else follows) likes an average RMS of between -23dB and -18dB.

There are two ways of achieving this in Audacity.

Method 1: the **Loudness Normalization** tool. You can find it under *Effect > Volume and Compression > Loudness Normalization*

Method 2: **RMS Normalize** is a third-party Nyquist plug-in tool created to do all the normalisation required of Quality Assurance to ACX levels. This is a tool that was created before Audacity added the Loudness Normalization feature to its latest versions. Many sound editors prefer to use this as it was designed especially for audiobooks. I prefer this method.

You could tweak these settings manually with the native Normalize and a few other tools that come with Audacity, or you can install this third-party plug-in which does it perfectly for you every time. You will need to download and install "RMS-Normalise.ny" separately from here: https://forum.audacityteam.org/t/rms-normalize/45334

Once you've done that (and restart Audacity), you'll find it under the Effects menu:

Effects -> RMS Normalize...

Select your entire sound file and apply RMS Normalize...

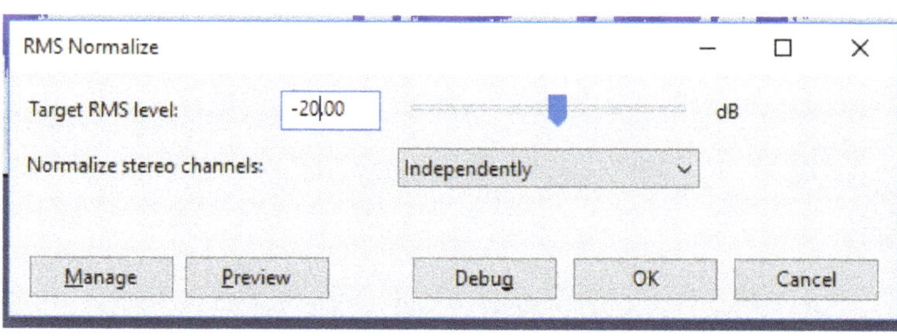

Figure 19: the RMS Normalise tool (Nyquist plug-in)

You can set the Target RMS level to -19dB or -20dB. Just avoid -18dB and -23dB, as these are the extreme limits of what's preferred, and you risk issues in QA if you're too close to the edge of what they want.

3. Limiter.

After normalising RMS, we'll find that our amplitude peaks have gone off the scale. To control these in a simple way, I apply a filter called a limiter, that will curtail those peaks without affecting the sound quality of the file. This effect comes with Audacity and is found under the Effects menu.

Effects -> Limiter...

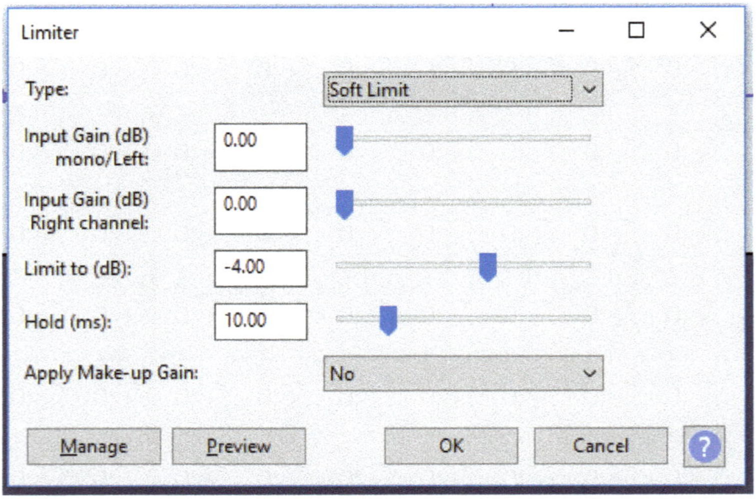

Figure 20: the Limiter tool, set to limit to -4dB in volume.

I prefer a soft limit and I set my limit to -4dB. Experiment if you want.

And that's all I do. I've mastered my track and am now ready to see if it'll pass ACX Check.

4. Quality Assurance: ACX Check

Some brilliant souls have created a Nyquist plug-in called *ACX Check* especially for audiobook creators. This analyses all your technical levels to see that they meet ACX standards. It doesn't come standard with Audacity and is one of the Nyquist plug-ins you'll have to download and install it from here:
https://wiki.audacityteam.org/w/images/5/53/Acx-check.ny

Nyquist plug-ins allow Audacity to be configured for many purposes, like checking for ACX compliance. Alas, Adobe Audition cannot use Nyquist plug-ins. Many users who otherwise love Adobe Audition will do all their editing in there, then export their file to Audacity purely to check for ACX compatibility.

This plug-in is useful for general analysis of your file, as well as ensuring you've hit your QA requirements. This plug-in is one of the big reasons I'm so fond of doing my audiobook editing in Audacity.

Want to find out the sound floor of a raw file before you've edited or mastered it? Use ACX Check.

Analyze -> ACX Check

First, select your entire file. Open the *ACX Check* analysis tool. It will analyse your file and let you know what the levels are, and if they would pass ACX Quality Assurance (in theory). This only measures the technical side of your file, not the human quality side.

To show you how it looks, I ran this over two sound files I had.

This file passed ACX Check:

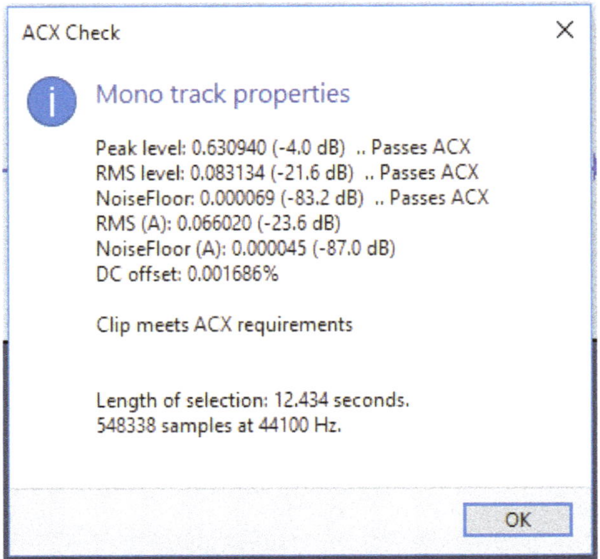

Figure 21: Passing ACX check.

This file did not, and ACX Check tells me why:

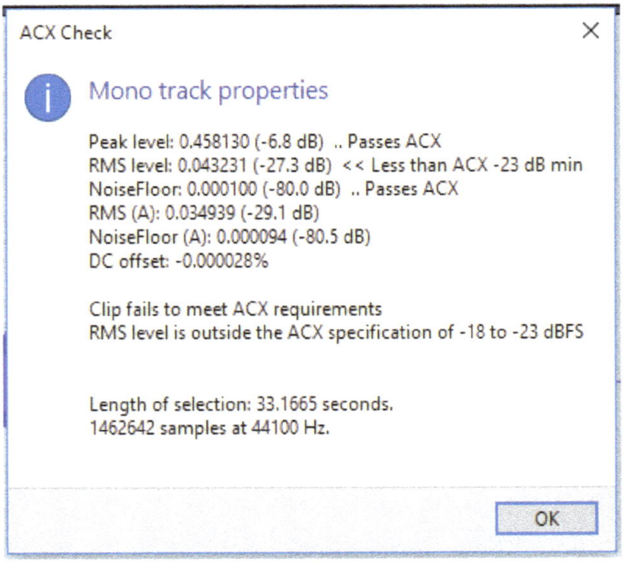

Figure 22: Failed ACX check for RMS levels too low.

If I fail the ACX check analysis, I'll review the results to see where my mastering went wrong.

- Noise floor fail: Did I run Noise reduction, or do I need to change my Noise reduction levels? Should I have filtered for rumble?
- RMS fail: did I run RMS Normalize? Did I do my mastering steps out of order?
- Peak level fail: Did I run the limiter? Did I do my mastering steps out of order?

If I don't pass the ACX check, I undo all my mastering (Ctrl-Z or Edit -> Undo [step], repeat until undone to error point) and try again.

Once a file has been mastered, I'll export it as an .mp3, and it's ready for upload. More on exporting to different file formats in a later chapter.

Noise Gating: Yea or Nay?

Sometimes certain generic or made-for-purpose mics (like a head mic) have onboard noise gating software. A noise gate is a filter designed to reduce the amount of noise in between narration phrases when laying down soundtrack. They're good for cutting out background during a phone call, but not so good for audiobook narration.

Noise gate is a Nyquist plug-in available in Audacity to control the levels of quiet between your narration. It opens to let through sound that meets its threshold and closes to cut off sound below its threshold. If you wish to try it, you can download it here: https://plugins.audacityteam.org/Nyquist-plugins/effect-plugins/dynamics-processing

If your room tone is quite loud and you can't get your noise floor to drop despite your best efforts to control your recording environment, a noise gate might help you.

It's a fiddly tool that, if configured correctly, can improve a raw recording and reduce editing time. However, getting it correct is a bit of a pain. An improperly configured gate will cause quality issues at the beginning and end of your narration phrases that will get your QA knocked back.

I do not use a noise gate. My room tone is quiet enough that it's not necessary and gating doesn't clean up my non-verbal sounds like a deep breath, a cough or a water sip. I'll still have to go through and clean that up myself manually.

Feel free to download the Noise Gate plug-in and play with it, if you think you need it. Much tweaking may be required in adjusting the threshold levels, but if you get it right, it could be a useful tool. Maybe.

Homework

Download this file:
Master_Sample_raw_track.wav from https://heidikneale.com.au/AGRYOA/ or see page 183 for link.

This file is raw, unedited and pretty much as basic as it gets. I recorded it not in my fancy studio with my high-quality condenser mic, but at my kitchen table with the on-board mic on my laptop. I've got a ticking clock in the background and I've got the A/C on.

It's got a few other issues like a too-low RMS level, but mastering will sort those out.

Apply the mastering steps in order to see if you can get that raw file to pass ACX check.

This is the ACX Check of the raw file before mastering:

```
Peak level: 0.998473 (-0.0 dB) << Exceeds ACX -3 dB max
RMS level: 0.059404 (-24.5 dB) << Less than ACX -23 dB min
NoiseFloor: 0.000245 (-72.2 dB) .. Passes ACX
RMS (A): 0.051564 (-25.8 dB)
NoiseFloor (A): 0.000152 (-76.4 dB)
DC offset: 0.000074%

Clip fails to meet ACX requirements
Peak exceeds ACX specification of -3 dBFS
RMS level is outside the ACX specification of -18 to -23 dBFS
```

Homework, part deux:

Mastering a file of your own.

Take one of the sound files you recorded earlier. Export the file as a .WAV file for a backup, in case you make a mistake in mastering and want to try again.

Give it an edit if you want, to remove unwanted noise and tighten up any mistakes or repeats you made. Remember to leave several seconds of room tone in your piece at the beginning and the end, ACX Check needs this for proper analysis. I recommend at least one second at the beginning and five seconds at the end of your file.

Analyse it with the ACX Check tool before you master it. How is your noise floor? What else did ACX Check pick up?

Master your file with the steps as listed above. If you want, you can run ACX Check after each mastering step to see how each step affects the file's quality.

After you have completed mastering your file, see if it passes ACX Check. Remember to save a backup of your original file in case you need to try again.

P.S.: Yes, that sound file from homework part 1 can be made to pass ACX:

Peak level: 0.630958 (-4.0 dB) .. Passes ACX
RMS level: 0.105189 (-19.6 dB) .. Passes ACX
NoiseFloor: 0.000305 (-70.3 dB) .. Passes ACX
RMS (A): 0.085692 (-21.3 dB)
NoiseFloor (A): 0.000169 (-75.4 dB)
DC offset: -0.000105%

Clip meets ACX requirements

Here's the edited and mastered file:

Listen to sound file:
Master_Sample_mastered_track.mp3 (see page 183 for link, or visit https://heidikneale.com.au/AGRYOA/)

How does it sound compared to the original file?

4.2 Exporting your Audacity project as an MP3

Here's a quick tutorial in exporting your project in the MP3 format, ready to upload to your distributor.

Once all the scenes of a chapter have been combined into a single Project and you've mastered them all so it passes *ACX Check*, it's time to export it.

1. Save your Audacity Project. You'd be surprised how often people forget to save their work before doing something major. By saving at key points, if you irrevocably mess up a step, you can revert to a previous saved version.

2. Go File -> Export -> Export as MP3... this will open a dialogue window.

3. Browse to the destination folder for your final MP3 files. I recommend creating one folder just for your final MP3s. This is considered good file management.

4. Give your file its name: For me, I choose to follow INaudio's preferred naming convention. Thus, the file I'm saving now is 001_Pride_Prejudice_Chapter_01.mp3.

5. Format conventions

- Bit Rate Mode: Constant
- Quality 192 kbps. You could go higher, but this will result in a larger file size.
- Channel mode: I tend to do all my audiobooks in mono, so I tick "Force export to mono".

Audacity will remember these settings, but it never hurts to check them every time. If different files within the audiobook don't match in format, you can get knocked back for Quality Assurance.

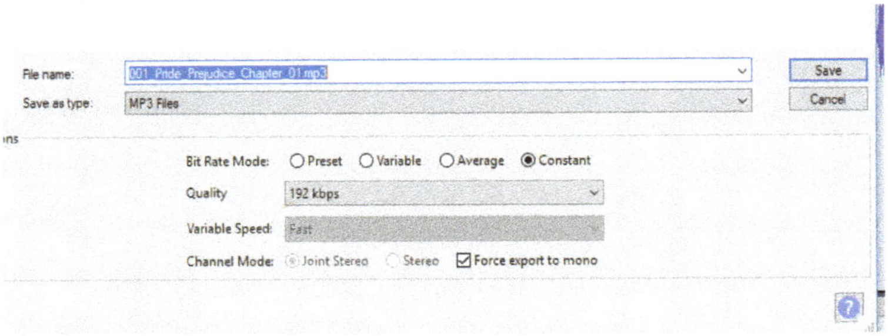

Figure 21: The format options in the Save window.

6. When everything is configured, click Save. Another dialogue window will open where you can edit the metadata tags.

7. Metadata tags

You have the opportunity to tag your audio files. I strongly recommend you do this.

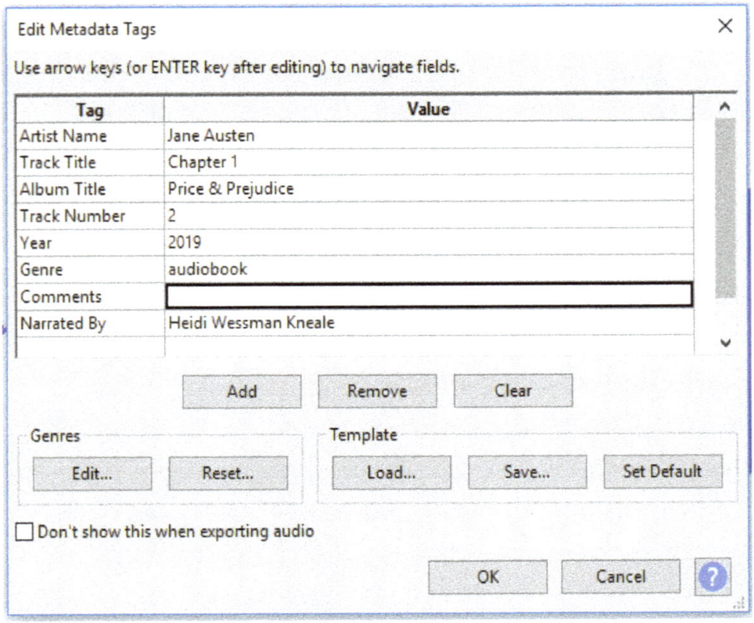

Figure 22: The metadata window.

Metadata tags to include:

- Artist Name: The author of the book
- Track Title: Whatever track this is, whether it's Opening Credits, Chapter One or Appendix
- Album Title: The title of the book
- Track Number: What order are the tracks in? Opening Credits will always be Track 1.
- Year: The year the audiobook is produced.
- Genre: audiobook
- Comments: Whatever you see fit. I tend to leave this blank.

Audacity doesn't default a narrator tag. Go ahead and create one by click the Add button. This will give you a blank line. Put in "Narrated By" and fill in the Narrator's name.

If you need further metadata tags, you can add them. Composer credit? Producer credit? Whatever you feel you need.

8. Creating a Template.

Since you're going to be creating a good dozen or so MP3s with nearly identical metadata info, save yourself some time and create a template. Keep the variable tags like Track Title and Track Number blank but all the static info should be filled in. Click Save... A new dialogue window will open. Don't change the file location. Give your template a memorable name and click Save.

Now, every time you create a new MP3 for this audiobook, you can load your template and have it pre-populate your metadata tags.

9. Export time.

Once your metadata tags are populated, click Okay. Audacity will now export your file in the .mp3 format.

Name	Type	Size	Title	Album	Length
000_Pride_Prejudice.mp3	MP3 File	6,837 KB	Pride & Prejudic sample rea...		00:04:51
001_Pride_Prejudice_Chapter_01.mp3	MP3 File	6,837 KB	Chapter 1	Price & Prejudice	00:04:51

Figure 23: final .mp3 files displaying metadata tags

Complete the export steps for all your chapters, including your opening and closing credits.

10. One Final Listen

It never hurts to do one last quality control check on your work. Listen to your audiobook mp3 from beginning to end. This is your last chance to catch any mistakes. Re-export your fixed file before you upload it.

Also, have someone else give it a listen. If you're the narrator/producer, you're probably getting sick of hearing this book. Your brain might be automatically making corrections or omitting errors. By having a fresh pair of ears listen, you can catch any mistakes your target audience will also catch.

Done!

Once complete, these files are ready to be uploaded to the distributor of your choice.

4.3 Overview of Post-Production

Industry Standards for Audiobook Quality

The industry standards for audiobook quality are as follows.

All files must be:

- Recorded in a human voice[1]. No computer text-to-speech narration. (Even Stephen Hawking's books had to be narrated by a live human. That's how strict the aggregators were.)
- Sampling rate of 44.1 kHz (remember that 44100 Hz wave sample rate we talked about when we started recording? That's what this is.)
- Channel format: either all files are in mono or all stereo. (A few distributors will even balk at stereo, preferring mono files only.)
- Every section must have a chapter/section announcement. e.g.: "Chapter Five". If it's in the manuscript, it needs to be in the sound file.
- Every book must have an opening credit track and a closing credit track.

---

[1]*With the onset of AI narration intruding into the audiobook market, some aggregators are accepting audiobooks created by AI. While some aggregators will allow AI-generated audiobooks, know that their inclusion is greatly controlled, and many markets still refuse AI-generated books. Plus many readers openly express a preference for the warmth and humanity of the human voice.*

- No extra (unwanted) sounds or outtakes. Edit out all the coughs, dog barks, passing trains and "let's try that again" outtakes
- RMS between -23dB and -18dB
- Peak amplitude below -3dB
- Maximum -60dB floor.
- Room tone: between 0.5 and 1.0 seconds at the beginning and between 1.0 and 5.0 seconds at the end.
- Maximum file length of 120 minutes. (Many distributors prefer shorter if possible. For example, Author's Republic insists that each file be no larger than 170MB in size. Depending on how you recorded your file, this may be well under the 120-minute limit.)
- Files must be saved in MP3 file format. (Some distributors may accept other file formats; INaudio will accept FLAC encoding. Stick with what you know.)
- Constant bitrate of 192 kbps or higher.
- Naming conventions must match your distributor's preferred standards.

Whew!

Don't feel that this list is overwhelming. I'm going to break it down so you know where and when to tweak things to ensure your soundtracks meet industry standards.

1. Human voice: That's either your voice in "Narrated by the Author" or a narrator you've contracted or the narrator provided to you by ACX or elsewhere. Why a human voice? Because computer text-to-speech synthesisers simply do not have the emotional depth, warmth, expression or liveliness that a human voice has. Narration is a performance and it needs to sound real. After all, your listeners are going to be hearing this voice for hours. A human brain will start to tune out to a computer voice after about ten minutes. You want to keep your audience engaged and a human voice is the best tool for the job.

"But what about AI?" you may ask? And fair enough. AI audiobooks didn't exist when I started producing audiobooks, but in the past couple of years, the technology has shoved its way into the market. At the time of print for this edition, most distributors refuse to let AI-narrated books into the market, but this may change and it may change fast. I touch upon this in a later section of this book.

Before I Record...

2. Sampling rate of 44.1 kHz. This is set before you start recording. There's a dropdown box in the lower left side of the Audacity window that allows you to set the sampling rate for the projects. Audacity tends to remember this setting. Also, this is an industry standard not just for audiobook but for music recordings and more. It is the optimal rate to ensure the maximum fidelity to the analogue signal with the smallest possible file size without losing sound integrity.

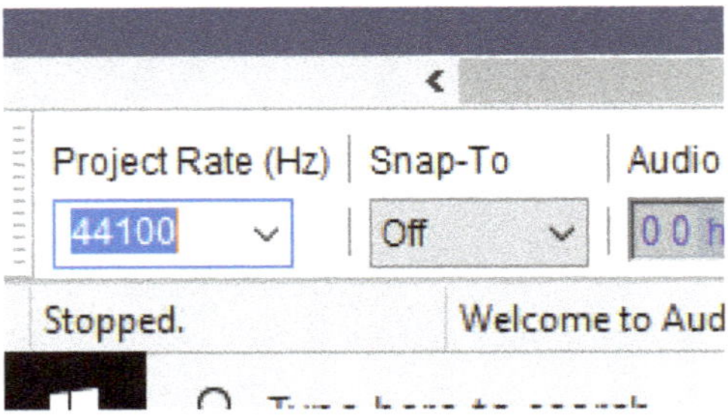

Figure 24: sampling rate set to 44100, just a smidgen over the top of human hearing range.

3. Channel format. Either set your microphone to mono input in the Device toolbar before you record, or you can mix your file down after you've recorded by going Tracks -> Mix -> Mix stereo down to mono. Note: you will need to have your whole track selected to do this.

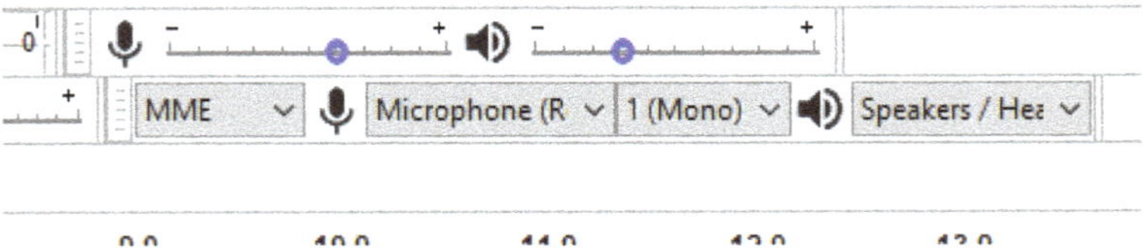

Figure 25: Mic set to Mono input.

As I Record...

4. Chapter announcement. I prefer to record each section as a separate project. I tend to have an average of three sections per chapter. When I begin a new chapter, I'll give the chapter heading. Have a listen to this sample from my book "Her Endearing Young Charms" (available where all good audiobooks are sold). Plain, simple, nothing fancy.

Listen to sound file: Chapter header sample.mp3
(see page 183 for link, or visit https://heidikneale.com.au/AGRYOA/)

5. Opening and closing credit tracks. You will need to record separate tracks for your opening credit and your closing credit. These are kept very simple, but have a minimal requirement:

Opening credit track:

Title of Novel
Subtitle (if present)
Written by Author Name
Narrated by Narrator Name

Distributors like us to keep the opening credit track as simple as we can. Anything else, like a Foreward, belongs in its own track.

Here's what I did for my novel "Her Endearing Young Charms":

Listen to sound file:
000_Her_Endearing_Young_Charms.mp3 (see page 183 for link, or visit https://heidikneale.com.au/AGRYOA/)

Closing credit track can be as little as:

The End

Okay, that's rather sparse. But that's the minimum required, as all the other valuable info like title, author, etc were included in the opening credit track.

There is one cheat for this: including the words "The End" as the last line in the last scene of the last chapter of your book. The statement "The End" is a requirement to let the listener know this is the end of the book. It doesn't matter if it happens at the end of the final chapter or at the beginning of the closing credit track.

To end a story with nothing more than "The End" is rather abrupt and not to our advantage as authors.

INaudio recommends adding a lot more info, thus:

This has been [Title], Written by [Author Name], Narrated by [Narrator Name], Copyright [Year of Manuscript and Name of Rights Holder], Production Copyright [Year of Audiobook Production] by [Rights Holder].

Here's my closing credit track for Her Endearing Young Charms. Note I include a little extra.

Listen to sound file:
019_Her_Endearing_Young_Charms_End_Credits.mp3
(see page 183 for link, or visit https://heidikneale.com.au/AGRYOA/)

If you use music as intro/outro, as I do, you will need to include music credit as well. (See Chapter 5.2 regarding music usage, as there are guidelines to that.)

You can include a few things beyond basics, but this is not the place to throw in a whole lot of unnecessary stuff. KISS principle applies.

What about backmatter? Backmatter is an invaluable marketing tool in an ebook. But does it work the same for an audiobook? Can it, should it, be included in an audiobook?

Very good question. Some say nay, some say yea. INaudio does allow backmatter (Appendices, endnotes, bibliographies, whatever), but it will be placed before your end credits. Definitely check with your distributor to see what their policy is on backmatter. For ACX, if you include it, they like any backmatter to match word-for-word with what's in your current version of ebook for Whispersync reasons. If you don't include it, they're okay, as long as the main body of book text is a complete match.

I tend not to include backmatter, and my end credits are pretty much all I do outside the story.

Your opening credit track is the absolute first track of your audiobook. Your closing credit track is the absolute last track of your audiobook. Always.

6. No unwanted sounds or outtakes. Best not to let them creep in when I record. Thus, the silent room in which I record. But if/when they do, I edit them out. I make the file as clean as I can, as unwanted sounds will distract the listener. Our goal is to give the listener a total immersive experience. We want them lost in the story, and not jerked out of it by mistakes we've made or strange background noises.

After I Record...

7. 8. 9. Mastering. Okay, the techie stuff:

- RMS between -23dB and -18dB
- Peak amplitude below -3dB
- Maximum -60dB floor.

We want every scene, every chapter to have the same RMS, peak amplitudes and noise floors. This will ensure consistency as listeners move from scene to scene, chapter to chapter.

Check out the Mastering section in Chapter 3.3 for step-by-step instructions on mastering your files.

10. Room tone. After I master my file, I do one last trim of the beginning and ending of my file to ensure one second of room tone at the beginning and three seconds of room tone at the end.

11. Maximum file length. Industry standards demand files that are less than 120 minutes in play length. This is approximately 12,000 or so words. Unless you're a long-winded author with extremely long chapters, file length probably isn't an issue. But if you finish mastering your file and it is long, you can divide it into two parts. Just make sure you have your chapter intro at the beginning of the file, e.g.: "Chapter Ten, continued." (See step 4.)

If, when recording, you suspect a file might be too long, you can do your file division while everything's still raw. Otherwise, you can do it now, after mastering. Just cut and paste your file into a new Audacity Project. Add clean room tone at the beginnings and ends of your two new files.

Time to save. I've posted a separate section for this, but here's the basics:

12. Format. Files must be saved as .mp3 format. Files must be a constant 192kbps or higher.

MP3 is a file compression algorithm that condenses your file down to a manageable size. It is lossy, meaning some information is lost in favour of compression. This is why this is the absolute last step in creating your audiobook file.

Industry standards require certain qualities to this .mp3 file, because once your distributors get their hands on it, they will be converting it again to different formats. Having a constant bitrate of 192 kilobytes per second (kbps) or higher, you minimise the lossiness.

To save your file as a .mp3 in Audacity, go File -> Export -> Export as MP3. (Note: Recent versions of Audacity include the LAME MP3 encoder to convert sound to this common format, so you don't need to install it separately. If you are using an older version, you may need to install the LAME encoder to export MP3 files.)

13. Naming conventions depend on your distributor. I go through INaudio, so they prefer the following naming convention:

000_Book_Title.mp3 (These are the opening credits, but don't get marked "_Opening_Credits.mp3". The title of the novel alone and the triple zero prefix is sufficient for marking it as the first file.)

001_Book_Title_Chapter_1.mp3

002_Book_Title_Chapter_2.mp3

003_Book_Title_Back_Matter.mp3

004_Book_Title_End_Credits.mp3

The tracks are given a three-digit number designating the order in which they appear. Next is the book title. Last is the personal name of the track

The first file 000 is always your opening credits track. Your last file is always your end credits.

And that's how you meet the industry standard requirements for audiobook narration.

Want to know more about cranking out high-quality files that pass QA? Here's some links to various distributors' quality requirements:

- ACX: https://www.acx.com/help/acx-audio-submission-requirements/201456300
- Author's Republic: https://www.authorsrepublic.com/creation#spec
- INaudio: https://voices.inaudio.com/technical-requirements-for-assets
- INaudio has a YouTube video for mastering in Adobe Audition: https://www.youtube.com/watch?v=x387JWPhKEg&t=452s)

Chapter 5 - The Extra Bits

In this chapter I cover the final steps of audiobook production. I'll touch on audiobook covers, info regarding including music in an audiobook, and distribution. I'll also add bonus info on outsourced narrators. Sometimes authors will give recording an audiobook a go, realise what's involved, and opt to outsource to another narrator. That's okay. I think that authors who've given the process a go are more likely to be successful when working with external narrators. You know what's involved in the process, you know what's possible, and what to expect.

Ultimately, the purpose of this book was to give you an inside look into what it takes to produce an audiobook, and how I put those skills to work creating my own.

Maybe you have tried your hand at some of the exercises I included to see for yourself what producing an audiobook is like. I hope you have an opportunity to lay down some soundtrack, edit, master it and produce a product you're happy with. If not, that's okay. It's not an easy process and can take much time and focus. Producing your own audiobooks isn't for everyone, but it is worth it to give a small project a go, if only to have a better understanding of the process.

4.1 You as Narrator

A narrator is a must for an audiobook. Text-to-speech generation is not permitted. You need a live human being. While the spectre of generative AI is pushing its way into the space, most markets still insist on the human voice.

So find that closet. Hook up that mic. Configure your inputs. Pull out your manuscript.

Before you start working on a project in earnest, I recommend a sample run or three. With previous homework I recommended you record you reading some of your text for a few minutes. Now, we're going to level up.

Homework: **Test Run.** Do a test run of you reading your text for thirty minutes. Yep, that's a half hour. I recommend a ten-minute warmup before you start recording. Then hit record and go for it. Set a timer to let you know when your half hour is up. If your alarm goes off while you're reading, don't worry. This isn't your final sound track.

Here's a checklist of things to consider.

Pre-test run:	
✓	are you in a quiet space with good room tone?
✓	are you wearing soft, quiet clothes? Certain textiles, like nylon, can be noisy if you move around. Flannel, fleece, soft cotton, these are quieter clothing and will not leave unwanted noise on your recording. Likewise, avoid jewellery and watches, as these can also be noisy.
✓	do you have cup or bottle of water next to you? Narrating will dry your mouth and throat. You will want to wet your whistle frequently. NO carbonated drinks, as they will lead to burps and your mic might pick up the carbonation. NO milk, as this will create mucus in your throat which can interfere with narration.

During test run:	
✓	is your microphone in optimal position? Do you need a pop filter?
✓	is your noise floor as low as you can get it without compromising your voice levels?
✓	is your narration volume showing up on the sound level meter as rich and warm?
✓	are you reading at a good pace, not too slow, not too fast?
✓	are you comfortable? Could you sit like that for an hour?

Post-test run:	
✓	how do you feel? Narrating can be hard work, as you're working muscles you might not use that often. If you're narrating an entire novel, you may be looking at a good fifteen or twenty hours or so of narration. This is a marathon. Don't be surprised if you get tired.
✓	how's your throat? Are you sounding hoarse? If so, you aren't positioning your voice properly. Refresh the section on vocal positioning in our Narrating chapter, look up a few YouTube classes.
✓	listen to your test run. What do you notice? What would you like to change? What do you like best? Does your pacing speed up, or were you able to maintain a steady pace?
✓	have someone else listen to your narration and give you honest feedback. Do you sound too flat? Too dramatic? Do your words flow well, or do you have some issues with articulation?

A test run will give you a good idea of what the challenges will be facing you when you narrate. Some of these things you will be able to fix. Fix them and try for another half-hour test run.

If you're pretty satisfied with your half-hour test run, if you've worked out your bugs, your foibles, your little issues that can't easily be fixed in editing, go ahead and record your whole project. I recommend doing it in several sessions. If you make mistakes, don't give up and think the whole session is ruined. Take a breath, redo the sentence (or paragraph if you must), and carry on.

Don't worry if you can only fit in a fifteen-minute session a day at the beginning. If you start feeling tired or you start sounding hoarse, it's okay to stop and give yourself a rest. This is not one of those things you can push through when you start flagging. Your tiredness will show in the quality of your voice. As our aim is to provide a product with consistent quality, obey this bottom line and stop if you're getting tired.

Do make some time every day to narrate. You will get your project done sooner. This regular practice will work your muscles and you will get better. Narration sessions close together will also contribute to the consistency of your project quality.

How long will it take to create an audiobook?

It can take between five to eight hours of recording, editing and mastering to create one finished hour of professional-grade audiobook. The better you get at the process, the faster it will be.

Currently, my ratio is about 5:1 work hours to finished hour, with about 9000 words per hour. This gives a 90,000-word book about ten finished hours in audiobook format.

Your mileage will vary.

5.2 Music

You may have noticed many audiobooks with music accompaniment as the intro/outro to a chapter, or to mark a change in scenes. I use music in all my audiobooks, even if only to justify years of music lessons and my Bachelor's degree.

Can you have music? Sure. Just be sure you have the right to use the music.

The music industry is very strict when it comes to music rights. They are not afraid to go after someone using music improperly. They are vicious, rabid weasels and rightfully so. We respect them for it.

But don't let this scare you away from using any music at all. You don't have to have music of any kind in an audiobook. Many audiobooks are "dry" aka without any music or anything else other than pure narration. That's perfectly acceptable.)

Music done well can greatly enhance an audiobook. Done poorly, it can cause issues. Also, listeners have their preferences. Some have loved the music I've included in projects. Others dislike it.

Including music is entirely up to you.

Music should not overwhelm the narration. If I have music overlapping narration, I make sure the music does not peak more than -20dB to -18dB.

In Audacity, music gets its own track. Master the vocal track separate, then add music, adjusting the music track so it doesn't overpower the vocals.

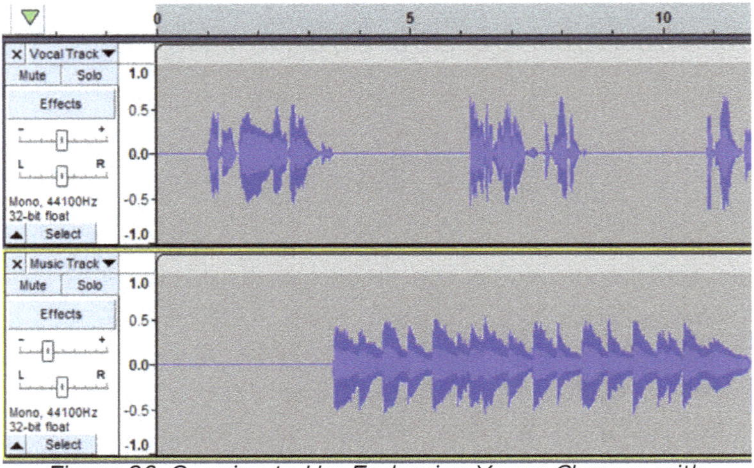

Figure 26: Opening to Her Endearing Young Charms, with vocal track on top, and music track on bottom.

Here's my narration (top track, aka "vocal track") with the music track underneath. Both have been recorded in mono. Note that the volume (under the Effects button) is turned down for the music track in that track's control panel.

Why Include Music?

For me, including music can help set the mood or lift the emotional resonance of a book. For Her Endearing Young Charms I chose a simple piece played on the harpsichord to evoke the Regency Era. For Richer For Poorer's music has a wistful-yet-hopeful oboe playing the main theme.

Sourcing Music

Me, I write my own, being a musician. I own all the rights and I can tailor the piece to fit my chapters. If you are also musically enabled, feel free to make your own music. Otherwise, there are other sources where you can obtain music.

21st Century technology has given us the Digital Audio Workstation (DAW). This is versatile software that lets you create music. I use Mixcraft at home and LogicPro when at university. If you have a background in music, and have access to a DAW, feel free to play with it and see what you can create.

If you can play a musical instrument (piano, guitar, whatever) and you're game to experiment, feel free to record yourself in Audacity and layer that if you want. Most people, however, aren't game to record their own, and that's fine.

This is the theme music I wrote for Her Endearing Young Charms. I designed it so I could loop it, make it as long or as short as I wanted to, and generally play with it.

Listen to sound file: Theme_music_HEYC.mp3

(see page 183 for link, or visit https://heidikneale.com.au/AGRYOA/)

Because this was a full-length audiobook, I wanted a little variety in the music. At the end of the second act, bad stuff happens, so I wanted some music to lend some atmosphere to that chapter.

Listen to sound file: Merrybelles_Lament 2.mp3

This is the theme music for For Richer, For Poorer:

Listen to sound file: FRFP_theme_music.mp3

I put some thought into the mood and theme of the novels I narrate and do my best to find music that matches, whether I compose it myself or source it elsewhere. You may want to put a harpsichord to a Regency Romance but for a contemporary romance novel about rock stars, you may want some groovy beats that slap instead.

Custom Composed Music

Know a composer who's willing to write you a few seconds of music? As long as you can get a contract stating the usage terms, this is a possible solution. Fivrr has composers who can crank out a few seconds of ditties for you, and ensure you've got the rights for its use. Work-for-hire is a good way to go, as it means you own the copyright.

When contracting a composer for audiobook music, you want the following products, at a minimum:

- Title track: This is the "theme music" that will overlay your intro (opening credits) and your outro (closing credits). Something that thematically matches your story is best, with the potential for looping. This music tends to be 20-30 seconds long
- Entracte tracks: these are 2-4 very short "in-between" music riffs, about 5-8 seconds long. This is the music that goes in between different scenes. It can be a melodic snippet from the title track, but doesn't have to be. Don't make these too long.

If you do work with a composer, GET A CONTRACT. Especially, make sure your contract states who owns which rights. If you don't own the rights, ensure the contract states you have been granted the rights to use this music for commercial purposes, especially for audiobooks. Maybe include a clause if using it for advertising as well? This is important if you are considering using the music as background in an Instagram, YouTube or TikTok promotional video.

Royalty-Free Music

You can find music online. Search for "royalty free music for audiobooks".

Royalty-free music means that you pay one license fee and the music's all yours to use in perpetuity. (Any music that's NOT royalty-free will incur a royalty charge for every audiobook sold. Don't go this path. It's just not worth it.)

Find a few sites and look at their licensing agreements BEFORE you listen to their samples. Some sites will have specific licensing for various usage. Check to see that they'll allow music usage for audiobooks. Some will, some won't.

- For example, Envato has lots of really nice music. https://1.envato.market/eNQkO These are their license terms. https://elements.envato.com/license-terms Obviously you will need to maintain a subscription to Envato until your audiobook is complete and ready to upload to maintain use of the piece. Once the book is done, you can unsubscribe.
- MediaMusicNow licenses music for audiobooks up to 5000 copies sold. https://www.mediamusicnow.co.uk/royalty-free-music.aspx Then they insist on a license extension: https://www.mediamusicnow.co.uk/help-support/answers/what-can-your-music-be-used-for/audio-books.aspx
- BenSound will NOT let you use their music for audiobooks. Sorry. https://www.bensound.com/licensing

- GettyImages also has stock music royalty free. What are their licensing requirements? Dunno. I was unable to find any information regarding this on their website. https://gettyimages.triplescoopmusic.com/

Royalty-free music is probably the easiest solution for sourcing music. You know exactly what rights you are purchasing and how to use it.

There are other sources of low-cost or free music available, but you will need to be savvier about their use.

Creative Commons Music

Yet another source is Creative Commons music and music recordings in the public domain. This is a good way of finding fee-free music.

Be careful about Creative Commons free music. A Creative Commons license will often state "non-commercial" use. As our audiobooks are commercial, this omits our using it. Check the license to ensure it doesn't restrict you to "non-commercial" use. If it's okay for commercial use, have at it and enjoy!

Public Domain Music

Public Domain Music is music that is not restricted by intellectual property rights. If the music isn't covered by a copyright, then you might be free to use it. Do your due diligence and make sure that whatever your using, and the version you're using isn't covered by a copyright. Sometimes the music score is public domain, but the sound recording of that score is not.

One possible free source for music is Musopen, https://musopen.org/ a site dedicated to copyright-free music. Check the licenses, as some of the music here may be under a "non-commercial" Creative Commons license. At least on Musopen they let you know exactly what license governs a piece.

Title	Composer	Form	License	Instruments
Gnossiennes no. 1	Erik Satie	Piano piece		Piano
Trois Gnossiennes	Erik Satie	Piano piece		Piano

Figure 27: A snippet from Musopen showing easy information about which licenses are available for which pieces.

The Gnossienne No 1 is copyright-free (Creative Commons 1.0 license)

The Trois Gnossiennes is non-commercial Creative Commons (CC non-commercial, no derivatives 3.0 license)

Just because something may be licensed under Creative Commons doesn't mean usage will be an unrestricted free-for-all. Worth checking what each license permits.

If you do wish to use some of Musopen's music, ensure that it's not a lossy version, or a version that's lower quality than the audiobook industry standard. This will affect the quality of your audiobook. See if you can acquire a lossless version of the music.

5.3 Audiobook Covers

Audiobook covers are traditionally square. This is a leftover from the days of CD cases, which are square. It's a deliberate design choice, as the square cover is a data point letting a potential purchaser know this is an audiobook and not something text-based (like an ebook or paperback).

Industry standard for audiobook covers are:

- File format: JPG, PNG, or TIFF only.
- Minimum size of 2400 x 2400 pixels.
- Minimum resolution: 72 dpi.
- Colour depth: 24-bit (True Color)
- RGB colour (not CYMK).

Images must be a true square, not a rectangular image with borders. (Because metadata of signalling this is an audiobook. Borders don't break up the "paperback" rectangle. If you can't get your cover artist to make an audiobook version of your cover, one thing you can do is add a bar, but fill it with information. Make it look like a deliberate design choice. Distributors also prefer that you clearly label the cover as an "Audiobook". This reduces confusion on marketing pages.

The audiobook cover doesn't have to match the ebook or the print cover, though you can if you want.

Examples of book covers vs audiobook covers

These are some of my book covers along with their audiobook counterparts:

5.4 Putting It All Together

Now that we've recorded, edited, mastered and exported our audiobook files, have sorted out the cover and anything else, now we upload to the distributor of our choice or package it for our own personal distribution.

The three primary audiobook distributors are:

- Audible/ACX https://www.acx.com/
- Voices by INaudio https://www.inaudio.com/for-authors
- Author's Republic https://www.authorsrepublic.com/

...and there are a few other niche ones you can look for. You don't have to go with the big three.

Also, you can individually upload to markets like

- Kobo Writing Life https://www.kobo.com/us/en/p/writinglife
- Google Play Books https://play.google.com/books/publish/.

Other distributors, like Apple Books (and iTunes), do not allow you to upload directly and you may have to go through an aggregator to reach this and other markets.

Originally, I was with Findaway Voices (now INaudio) because I am Australian. ACX doesn't want to work with Australians who don't have a US or UK bank account. Yes, while there are strange options and workaround one could do to trick ACX into dealing with me, the hassle of acquiring yet another foreign bank account purely for dealing with ACX wasn't worth it to me. (It might be for you; that is for you to judge as part of your author business plan.)

The main reasons I chose Findaway were that they're willing to do business with Australians and they distribute widely. Also, they're non-exclusive. However, they ONLY do distribution now. They do not handle production. (Once upon a time they did, but ever since being acquired by Spotify, they discontinued that offering.) Not a problem if you're making your own audiobooks, like me.

You will need your own ISBNs. While Audible and INaudio can provide them for you, they are only for use within their distribution. You cannot take your audiobook with this ISBN and distribute it elsewhere. (As an indie author, it's generally a good idea to have your own ISBNs anyway. Hit up Thorpe-Bowker in Australia for yours if you're Aussie.)

Audible/ACX tends to be American-centric, which I feel is rather nationalist of them, especially as they're selling to an international market, supposedly. You have to have a Canadian, Irish, UK, or US bank account to work with them (unless that's changed since the publication of this book). They are tied to Amazon and can be problematic if your title isn't available on their platform in ebook or print format. There are a few scandal rumours drifting about regarding book returns and low royalties. Plus they're exclusive. If you use their audiobook production, you are locked in to a seven-year contract.

Author's Republic tends to be friendly towards Australians but can be touch and go with distribution, access and a few other things, at random. Plus they don't always advertise their royalty rates.

All three distributors have their pros and cons. Research them all, see what they offer and feel free to choose the distributor that is best for you.

Naming Files

Nevertheless, whatever distributor you choose, they will have naming conventions and other requirements for the files before you can upload them.

For reference, INaudio's requirements are here: https://voices.inaudio.com/technical-requirements-for-assets

They prefer:
000_Book_Title_Opening_Credits.mp3
001_Book_Title_Chapter_01.mp3
002_Book_Title_Chapter_02.mp3

ACX isn't as stringent in its naming conventions. Their information is here: https://help.acx.com/s/article/what-are-the-acx-audio-submission-requirements

They prefer theirs to be:
00_OpeningCredits.mp3
01_Foreward.mp3
02_Chapter01.mp3
03_Chapter02.mp3

...and so on. Check with your distributor for their preferred naming conventions. The naming conventions are to ensure that your tracks end up in the correct order. Nothing's more frustrating to the listener to end up hearing Chapter 4 before they've heard Chapter 3.

The **metadata** of each .mp3 file also needs to be consistent. When you originally exported your .mp3 files, you had an opportunity to update the metadata.

Metadata serves several very useful purposes. Chances are your audiobook listeners are going to install them on a PC or an iPod or other listening device. This device will sort your audiobook according to track. It will differentiate them according to Album Title or Artist Name. It will display the Track Title to the listener. All these metatags ensure that the right chapters come at the right time and that they don't get mixed up with each other, or even with other sound files.

- Artist Name: The author of the book
- Track Title: Whatever track this is, whether it's Opening Credits, Chapter One or Appendix
- Album Title: The title of the book
- Track Number: What order are the tracks in? Opening Credits will always be Track 1.
- Year: The year the audiobook is produced.
- Genre: audiobook
- Comments: Whatever you see fit. I tend to leave this blank.

If you forgot, or you need to make changes to your metadata after you've exported your .mp3, you can update this after you've created the .mp3 by right-clicking on the file, selecting properties. A window will pop up. Select the Details tab and click on the fields you wish to edit.

Sample File

One more thing your distributor may ask for is a five-minute sample file. This can be the first five minutes of your Chapter One, or another passage from early on in your book. Each distributor will have their own requirement as to what they want. Some are okay with background music, others want pure narration. This sample file is used as a teaser or sample to potential buyers. Many people will choose audiobooks based on how the narrator sounds.

Here's the KoboBooks page for my audiobook "Her Endearing Young Charms".
https://www.kobo.com/au/en/audiobook/her-endearing-young-charms-2

The sample file I created for Findaway can be heard through the "Play Sample" link.

To create a sample file, open up a full chapter that's already been mastered in Audacity, then simply cut-n-paste a five-minute section into a new file. Save, export and name as per convention.

Packaging your Audiobook for distribution

If you're uploading your audiobook to a distributor like Audible or INaudio, you're pretty much done. All you need to do is upload your .mp3 files, your audiobook cover and that's it! They'll do the actual packaging and distribution.

But if you're packaging up your audiobook to distribute to readers via a different source (like, say, your website), the easiest method is to zip all your .mp3s along with your audiobook cover into a .zip file. This will provide a compact single file you can email or they can download. A .zip file is unzippable with pretty much any compression software like WinZip, 7-Zip, etc.

Once upzipped, the listeners can load the .mp3 files to the listening device of their choice.

If you're distributing via your personal e-commerce site, do you need a sample file? Absolutely! Giving your potential audience the ability to sample how the audiobook sounds is just as important as the blurb or book cover. Nothing sucks more than someone purchasing your book, then wanting to return it because they don't like your voice.

While I don't have a personal e-commerce site, I have linked my sample files to my audiobooks on my website: https://heidikneale.com.au/audiobooks/

5.5 Sourcing a Different Narrator

A narrator who's not you...?

So, you've gone through this book and faithfully completed your homework. You've made a small audiobook project (congratulations!!) and now have a good idea of what it takes to create your own audiobook from scratch.

By now you will have a good idea if you have the capacity to record your own audiobook. If you do, go forth and be mighty. I wish you great success.

But maybe you've discovered along the way that recording an audiobook isn't for you. Maybe after doing a few test runs, you realise you don't have the voice or the patience or the skillset to give your book the narration you feel it deserves. Don't feel bad if this is the case. Self-narration is not for everyone.

Does that mean it's the end of your journey? Not necessarily. Creating an audiobook isn't an all-or-nothing proposition. Maybe narrating isn't for you, but you're fine with editing and mastering. Good news, you can outsource narration.

What to look for in a good narrator

- **Good articulation**. You want a narrator who speaks clearly and is easily understood.
- **Pacing**. They have a sense of rhythm that fits the text and makes it flow well. They don't sound too fast or too slow.
- **Voicing**. Characters have different "voices". You can tell the difference between dialogue from Our Heroine and from Our Hero. A good narrator will give a certain inflection to each character, allowing the listener to tell them apart. This is more than simply pitching your voice high for Our Heroine and dropping it low for Our Hero. A good narrator will get a sense of the characters from your text and do their best to differentiate them in subtle ways.
  And it is subtle. You shouldn't get a male narrator squeaking out Our Heroine in a falsetto. An excellent example of a narrator doing an excellent job with character voices is Nathan Lowell and his "Golden Age of the Solar Clipper" series. He's got both male and female characters, and you never have any issues telling them apart. https://scribl.com/books/P2A75/quarter-share I highly recommend anyone wanting to hear good narration by an author to go listen to his work.
- **They 'get' the book.** You will be able to tell if they're getting the book within fifteen minutes. They'll have a sense of belonging, that they're into it and they're making it real. They're engaged and your words are brought to life through their voice.

- **Communication**. You're a team, when it comes to producing this audiobook. You and your narrator need to communicate. They'll have questions and you'll need to share your vision. It's not like you're going to toss them a copy of your manuscript and they're going to send back a perfectly packaged audiobook. They'll have questions about name pronunciations, characterisations, and more. They're asking questions to get a good sense of what you want. Work with them. Help them so they can give your book the performance it deserves.

Things You Can't Expect

- **Absolute perfection**. Remember, your narrator is human. Do not throw a fit and refuse to accept the whole book simply because they mispronounced one word out of a hundred thousand. (I have seen this happen.) Trust them in their craft that they'll do a good job.
  Your narrator is interpreting the words you've written. They're going to bring an interpretation to your work that might not exactly match the interpretation in your head. What you're looking for is Good Enough. Do they bring your story to life? Is their performance suitable to the text? Does their work meet Quality Assurance? Then it's good enough. Perfection is impossible. Suitability is possible.
- **Sounding like your main character.** They're not playing your characters. They're playing your narrator. A narrator is a character all his own, as he interprets your authorial voice. Stephen Fry

doesn't sound anything like Harry Potter, yet he narrated quite a few of Jo Rowling's novels successfully. Okay... or mostly successfully: https://www.youtube.com/watch?v=ysu7DYRgE5o&t=1176s

- **Cast of thousands**. Most audiobooks are narrated by a single author. Essentially, it is someone reading a book to you. Voice-casting different actors might make your book sound "real", but it will cost you money, as each actor needs to be paid for their time, and not just a Per Finished Hour rate. And are you going to cast just Our Hero and Our Heroine, or are you going to cast every character? What is your purpose for having different actors for each character? An audiobook is not a radio play or a movie. It's someone reading a story to you.
- **Royalty shares as payment**. There are many payment tiers a narrator/producer may accept. One of those is royalty shares from sales. Unless your book has a demonstrated strong sales record that will translate into strong audiobook sales, chances of you finding a decent narrator willing to work for royalty shares is almost zilch. Unless they're happy to volunteer their time (as narrators for Project Gutenberg would do), or they're new to audiobook narration and are trying to build up their portfolio, you're going to have a hard time attracting narrators willing to audition for your book if all you're offering are royalty shares.

What Services Do You Want From a Narrator

Plenty of narrators are out there, narrators who are more than happy to help you with this task. We thank you for giving an actor or narrator paid work. It's much appreciated.

Sourcing a narrator is not difficult if you know what you're looking for.

- Are you sourcing a voice talent only? If so, you will be expected to provide studio space for recording, and post-production.
- Will your narrator take care of the raw recording in the studio of their choosing? You handle the post-production.
- Will your narrator do the recording, editing and post-production, delivering to you a finished product ready to upload and distribute (aka "Ready to Go")?

Before you independently contract a narrator, you need to know how much work they will be putting in towards the completion of your audiobook. Ask them if they produce as well as narrate. Ask them what product they plan to deliver (raw recording or finished industry-standard MP3?) Not every audiobook narrator is also a producer. Best to know this before they dump fifteen hours of raw track on you.

So, why would you want to hire a voice talent for raw files? Cost. And that's it.

Hiring an actor to provide voice talent alone will cost you an average of AUD$30[1] or more per raw hour of their time.

---

[1] Payscale.com https://www.payscale.com/research/AU/All_Actors_%2F_Actresses/Hourly_Rate

This would not include the cost of studio time, editing, mastering or post. This is just for their acting skills alone. If it takes an actor about fifteen hours to record the raw, unedited soundtrack of your novel, then that would cost about AUD$450 minimum.

If they're a member of Actors Equity, SAG-AFTRA or another union, there will be specific guidelines covering their pay scale, minimum call-out hours and working conditions. If your narrator has an agent, that agent will be able to provide you with a rate card dictating minimum call-out times and working rates. Unionised and agented actor rates will be more than AUD$30/hr. Just be aware.

You don't have to go with an Equity or an agented narrator. If you could find someone from a local acting school who's willing to negotiate rates with you, that's fine. They'll probably come out cheaper, and they'll be happy to have some clips for their portfolio. Just be prepared to deal with a bit of inexperience, and recording may take a little longer than expected while they find their chops.

Regardless of who you source or from where, ALWAYS SIGN A CONTRACT. You are entering a work-for-hire situation and some sort of agreement is legally required. ACX has some nice boilerplates you can use. Actors Equity, SAG-AFTRA and other unions all over the world will have sample contracts you can look at for reference.

The contract will state what you expect from your narrator, their pay rates, their working conditions, including who's responsible for the recording studio, deliverables, deadlines, kill fees, etc.

If you're hiring voice talent only, the charges will be per raw hour of work put in. If you're hiring a narrator/producer, the charges will be per finished hour. This rate, while appearing higher than the raw rate, covers the recording, editing, mastering, post, etc. A good estimate to see how much you'd be paying per raw hour, divide this number by five.

Where to find narrators?

1. Source from Audiobook Production Companies

Some companies like ACX and Author's Republic will help you source narrators. They've got a whole stable of professionals who will give you back a professionally-finished audiobook, ready to go. This is probably the easiest (and most cost-effective) way of getting your audiobooks made.

ACX lets you browse their narrator database and you can approach a narrator you like. They also let narrators look through your posted projects to see if there's something they'd like to work on. Author's Republic also gets you access to their database of narrators.

ACX and Author's Republic narrators will deliver a finished product for you. That is one of the advantages of working with these companies.

You don't have to go through a distributor's stable of narrators. You can source your own, produce your own audiobooks and simply upload the final project to the distributor. This is how I upload my self-narrated books.

There are other audiobook production companies out there. Check their terms and their prices. I haven't researched them much, so I can't recommend or dis-recommend anyone.

2. Contract an independent narrator/producer

Simply Google (or Bing) "audiobook narrator" to get pages and pages of companies and independent audiobook narrators. Most narrators who work through these companies are narrator/producers and will charge you per finished hour. Depending on the company, some narrators may be willing to provide you with raw recordings instead for a lower hourly rate. This can be less expensive, assuming you don't have to contract out the editing/post-production as well.

3. Local acting/music school

If you're willing to take a chance on experience and quality, you could find a student and work with them. The good news is they will be in active training, will be familiar with their instrument (aka voice) and will be up for the task, especially as you're paying them. The bad news is that they'll only be a voice talent. Odds are they will have no production skills and the realm of editing/post will fall to you or someone else you hire. But as I have shown, if you have a good-quality raw track, editing, mastering and post-production aren't seriously challenging. Chances are you could do it yourself. More bad news: if your tracks do get knocked back in Quality Assurance by your distributor, you will need to hire out your actor again for that chapter. Hope they are available.

4. Hired Guns

Various work-for-hire sites like Fivrr and their mates have plenty of narrators hawking their wares. Be sure to ask them the right questions before contracting them. Be cautious, however, as these work-for-hire sites can be rife with scammers.

Narrator Pay Rates

For narrator/producers delivering a finished project, there will be various pay scales available to you, depending on who you source through.

Royalty Share. This is where the narrator gets half of all royalties made on this audiobook. That's their compensation. Very few companies offer this. ACX is one of them. Their contracts entitle the narrator to half the net royalties the audiobook earns for seven years (or whatever the term of their contract is).

While this option may appeal to the very budget-conscious author, be aware that unless your book is rip-roaring along in print and ebook sales, the chances of any narrator willing to work for Royalty Share is slim-to-none. If book sales are strong and your previous audiobook sales are also strong, a Royalty Share project will be more appealing to a narrator.

Half-n-Half. Some companies (usually small producers) will offer a deal where you pay half the narrator's rate up front and the other half of their compensation is to be made up in royalty share over the X number of years (or whatever the contract states). My publisher did this for a while.

Per Finished Hour. This is a work-for-hire rate. You pay the narrator to record, edit and post-produce the audiobook into a Ready to Go format meeting industry standards. All you need to do is upload the finished book to the distributor of your choice. Narrators do not collect any royalties on this deal. Rates tend to start at $125 PFH. $250 PFH is common and rates can go as high as $500 PFH. Occasionally you can find someone willing to work for less than $125 PFH, but these are rare. Usually they are either new to the industry and are building up their portfolio, or they're not sure how this show works.

Working with your Narrator

So you've chosen to outsource your narrator. How will you know they're the right one for your work? How would they handle your text?

Enter the narrator audition. This is where the narrator will give a five-to-fifteen minute reading of your text for your approval.

As you listen to their audition, look out for the following:

Technical quality

If you're auditioning for a finished-package narrator (who delivers your narration edited, mastered, and ready for distribution), their audition should match the industry standards for audiobooks.

Do they meet ACX standards? You can check this by downloading their audition .mp3, importing it into Audacity and running it through *ACX check*.

If you're looking for raw talent:

Is their audition free from unwanted noises (background intrusions, mouth noises, harsh breathing, outtakes, sibilance)?

Does their file sound overprocessed (sounds like they're narrating out of a tin can or wine glass)?

Are they narrating too fast or too slow? We're talking on the technical level. Does it sound like they want to rush through your text, or they're plodding through it?

Performance quality

This part is subject more to taste. Their audition tape might be technically acceptable, but how's their performance?

Do you like the sound of their voice? Did you want a male or a female voice? Warm or cool? Rumbly bass or crisp soprano? Think about what kind of voice your text deserves. If your text wants a female voice, you're probably not going to audition male narrators.

Does their voice fit the narrative? Is it working with the text? Does the voice feel like it fits the words?

How's their pacing? Does it match the narrative?

How did they handle character voices? You want to be able to tell the characters apart, but you don't want them to sound overdone or cartoony. (I think I'm guilty of doing this with Mrs Bennet's voice in our Pride & Prejudice sample.) A good actor will be able to give character inflection without needing to overdo it. Subtlety is the mark of a master of the craft.

The best thing you can do to see if a narrator is good at voicing characters is to give them a sample of your work from a scene that has a conversation between two or three characters. Can you hear the difference between the different characters when narrated?

A word on accents and dialects. Should your narrator use an accent or a dialect? Sometimes an author wants a narrator in an accent or dialect that matches the text. Regency Romance? Go sound British. Western Romance? Bring on that Texan.

Or not. Many narrators have a neutral accent that lends itself well to nearly any project. Neutral accents can be unobtrusive. Tell the story well, and your listeners won't notice you're an American living in Australia narrating a Regency Romance. (At least, that's the plan.)

When you look through various narrator profiles on ACX or other voice talent sites, they'll list the types of accents they can use.

Accents are a tricky thing to get right. It can take much training and practice to be able to pull off a convincing accent. Narrators will have their natural, or primary accent. For an American narrator, this may be a Midwestern Neutral. For a British narrator, this might be RP (Received Pronunciation).

I've known some highly-trained actors who are brilliant with accents, employing dozens as they see fit. Then there's me, who's terrible at accents. I can do a softer Midwestern and I can do an American Southern, because my family is from the South and I grew up listening to them speaking in their unashamed Mississippi drawls. Anything else just sounds fake from me, possibly because I can't reliably maintain it over the course of a project.

Self-narrators, unless you're hella good at accents, like, trained-good or you grew up with them, I recommend not attempting to put on an accent. Don't worry if your spoken voice's natural accent doesn't match perfectly to your ideal accent for your script. The authenticity of you as an author makes up for it because you're reading your own story to your listeners, not an actor doing the narration.

But if you're looking for a narrator who can do an accent, go for it. Ask them to perform their sample in the accent. Do you like it? Great. If not, this might not be the narrator for you.

When you listen to the auditions, you will know within the first thirty seconds if a narrator is definitely not for you. They just won't click.

Eventually, one will sound just right (or good enough). If you like them, select them and enjoy your beautifully narrated audiobook (in about 4-6 weeks' time).

Working with your Narrator

So, you decided you want someone else do to the narration? That's okay. Not every author is suited to providing quality narration, whether their voice isn't up to it, or their time is better spent, or they want a particular sound. There are plenty of narrators out there happy to collaborate with you on bringing an audiobook to fruition.

And it is a collaboration. You're not just going to hand your manuscript over to a narrator and magically four weeks later they return the perfect audiobook. Your input will be vital to your narrator's success.

Disclosure List

When you succeed in getting a narrator, you want to help them give the best delivery of your product. A disclosure list can help with that. This is a rundown of all the necessary information a narrator needs to know before narrating the book. Sure, a good narrator will read the book before they start their work, but they might miss out on something subtle. Your disclosure list ensures they won't miss out on vital detail.

A good disclosure list will contain the following:

- **All dialogue characters**, including a few characteristics for each one, especially the main characters. How do you envision them sounding? Do you mention anything about their voices in the text? What's their personality like? Overall, are they calm, nervous, angry, happy-go-lucky?

  An example of a character list:
    - Mr Bennet: middle aged man descended from gentility. Loves his daughters, barely tolerates his wife. Would like nothing better than a quiet afternoon at home.

    - Mrs Bennet: hypochondriac, always up in everyone's business. Would love nothing better than for her five daughters to marry and climb the social ladder. Shallow, with an opinion on everything.
    - Lizzy Bennet: practical, takes after her father. Endures her mother's in-your-face-ness. Thoughtful.
    - Kitty Bennet: adolescent daughter who hasn't quite left childhood yet.
- The **pronunciation of EVERY name** found in your text. You'd be surprised at how easy it is to mispronounce names. Give the good old Irish name Mahoney. Is it pronounced muh-hone-nee, or is it pronounced mah-nee? If you don't tell your narrator how every name is pronounced, even the ones you think might be obvious, you cannot get upset if they mispronounce a name.
- The **pronunciation of EVERY non-English word** found in your text.
- The **pronunciation of every technical term** (medical, scientific, etc). If it isn't standard, common English, clue your narrator in. When in doubt, clue your narrator in. On anything.
- **Default accent and every accent variation**. Now, most novels don't have accent variations, but some might have a Russian character, or a New Yorker, or a Mexican, or someone whose accent would be noticeable by its absence.

Your narrator may have a few other things they want to know. Ask them if there's anything else they need from you. Keep in regular communication with them throughout the project. Show you're open to questions and answer promptly. They may not be able to proceed in their work without info from you.

Don't get all up in their face, and don't insist on hearing their work in progress. They'll get it to you when it's production ready. Some narrators working under distributors may have a clause in their contract to get you Chapter 1 within a certain time period, the first half of the project at another deadline, and the remainder at the final deadline. It may be linked to a payment schedule.

There are good reasons for this. When you get that first chapter, listen carefully. If there's any issues (performance or technical) you feel need resolving, let the narrator know. That way they can resolve these issues before they commence the rest of the project. Nothing sucks worse than recording a whole project and then having the author not like an element you could have easily fixed early on.

It's okay if you have chosen not to do the whole project yourself. Creating an audiobook by your lonesome isn't for everyone.

Maybe you've decided to have someone else do the narration while you edit. Or maybe you're narrating and are outsourcing the editing and mastering elsewhere.

Anything you can do yourself (within reason) can help bring down the costs of an audiobook.

Chapter 6 - Legalities and Other Info

If you are planning on sharing your audiobook, whether with a few friends, as a giveaway to newsletter subscribers or releasing commercially to the world, you need to be aware of all the legal ramifications. Audiobooks are governed by rights, just as ebooks, print books, and more are governed by rights. Awareness of the legalities surrounding yours and others' intellectual property (IP) will help save you from potential future troubles.

I am not a lawyer. I cannot give you legal advice. But I can draw your attention to matters you need to be aware of and consider as you produce and distribute an audiobook. I do recommend you seek out legal advice if you have any questions about anything you may come across, especially if your production involves other people and/or services.

6.1 Legal stuff you need to know

Audiobooks are intellectual property. Always check if you have the right to record this book, even if you wrote it.

If you are the author of a book you want to turn into an audiobook, know that you originally own(ed) the audiobook rights. These are separate from the ebook rights, the print book rights, the movie rights, the video rights, the music rights... If you are NOT the author of the book you want to turn into an audiobook, find out who owns the rights, and if they're willing to let you use them.

Even if you've written the book, you still need to confirm audiobook rights are still in your hands. First question to ask: where did you get the book published?

If Indie, you most likely have audiobook rights.

If you went through a traditional publisher (whether the Big Five, a small press, or even a "hybrid" publisher), check your publishing contract. Some publishers will throw in a rights grab into their contract for stuff they have no intention of developing. All too often a fresh, starry-eyed first-time author might have signed a publishing contract without realising they were handing over rights they didn't need to hand over.

It's worth checking to see if your audiobook rights are tied up with a publisher. Please check this before you launch into any recording or production. I would hate to see you go through all this work only to have someone get nastily legal on you.

Originally, my Small Press publisher had claimed audiobook rights for several of my titles with the intention of releasing titles in this format. When they realised that audiobooks (on the whole) were not as profitable as they were hoping, they graciously reverted audiobook rights to any author who asked. For them, this was a brilliant move. Someone else can bear the burden and cost of production. Then, when the audiobook is out there, it helps drive sales to their ebooks and paperbacks.

Unless you're doing everything yourself (sound effects, music, cover, etc), know that anything you haven't created yourself is someone else's IP.

Don't breach someone else's rights. If you source music, sound effects, cover art, anything created by someone else, ensure you have the rights to use them in commercial audiobooks (even down to the fonts on the audiobook cover, if you're creating your own). Certain industries will come down HARD on rights breaches.

Always acknowledge the work of others.

For my audiobooks that were originally published by a Small Press, I did not have the rights to reuse their cover art, so I had to create new covers for them. No big deal. This is the way.

6.2 Contracts, Terms and Conditions...

... and Other Small Print.

If/when you work with other people, including distributors, always read the contract, the Terms & Conditions or anything else legal you need to agree to. It is in your best interest to know what you're signing.

Seek advice from a knowledgeable human being (several if you can) if you do not understand anything or have questions about anything in legal paperwork.

Several organisations like ALLi – the Alliance of Independent Authors (https://www.allianceindependentauthors.org/), The Australian Society of Authors (https://www.asauthors.org.au/) and more have services available to review legal contracts for members. And there are always law firms who handle contracts and IP who are happy to review a contract for you for a price.

If you have a literary agent, you should be running any and all contracts by her as standard operating procedure, even for work that she's not handling directly.

Keep copies of everything you agree to (including T&Cs) in an easily-accessible location. For example, make a copy of the Terms and Conditions from your distributor. Keep in mind that they tend to update these from time to time. It's a good idea to download and keep the latest version when you come across it. That way, if something happens, and you get locked out of your account (for whatever reason), you'll still have access to the information you may need to take further steps

---

When it comes to legalities, ensure you have a sufficient enough paper trail to cover your butt.

6.3 Generative AI

Prior to 2025, the use of Generative AI (GenAI) audiobook production was considered an industry taboo. It was natural human voice or nothing for the main audiobook distributors. Countless authors, like John Scalzi, have included clauses in their contracts that their books will only be narrated by humans. Part of it was that the quality wasn't as good as a warm human voice. Another element was the moral and ethical reasons. Nobody involved in the audiobook industry wanted to touch audiobooks generated by AI.

But in 2026, this began to change. Many companies, like ElevenLabs, are creating audiobooks with GenAI. Distributors like ACX and Spotify are starting to accept them, under certain conditions (including the tagging that the audiobook was produced with GenAI so listeners can self-select out of them if desired). They claim it's filling in a hole where human production can't keep up with the demand for audiobooks. Or so they say.

The moral and ethical reasons for the use of GenAI (or any AI, including LLMs) is complex and far too much for me to include here. But I encourage you to not plug your ears or stick your head in the sand over this. Learn as much as you can regarding the use of AI, as it is encroaching in creative spaces more and more. Are there AI tools that could make our creative lives easier? Yes. Are there AI tyrants that will be used to remove opportunities from creativity-producing humans? Also yes. The more we know about AI and its capabilities, the more we can control how it enters into our spaces and stand against intrusions that would threaten the basis of human creativity.

Will GenAI audiobook push out the existing market for audiobooks? Oddly enough, I don't think it will. While GenAI can create an audiobook in a fraction of the time it would take a human, it's not necessarily any cheaper to create that book. AI costs money to use.

Many listeners still have a preference for human voices and will actively seek out audiobooks created by human beings.

In spite of this cyber competitor, are you going to throw up your hands and not bother creating an audiobook?

No, I don't think you will.

One of the best things about humanity is this drive to create something new. If you didn't care about making your own audiobook, you wouldn't have picked up this book. But you do care.

So, go make yourself an audiobook, if you want.

I'd love to listen to it.

Chapter 7 - Let's Record an Audiobook!

You are persistent! If you've gotten this far and are still reading, you must really want to give making an audiobook a real good go. I'm not going to dissuade you at this point. If, after all I've told you, you still want to create an audiobook following the method I use to create an audiobook, do jump in with both feet.

I hope you've had a chance to...

- Select a short story or a chapter for your recording.
- Play around with Audacity
- Download the plug-ins to make mastering easier
- Play around with your voice and some of the acting exercises
- Find a good mic
- Do some test recordings
- Find a comfy spot in the closet (or sound booth or blanket fort) of your choice

...because we're ready to roll.

So, how do we put all this together? If you haven't already, go watch my YouTube video: "How I Record an Audiobook, from Go to Whoa!" showing the whole process. **Youtube.com:** https://www.youtube.com/watch?v=sFy8cDZ_XxU Watch me select my text, read and record it, give it several editing passes, clean it up, master it and then when it passes my personal Quality Assurance, export it as an .MP3 file, ready to upload to my distributor.

Now it's your turn.

Last bit of homework

Take that thirty-minute sound file from your last assignment. You've recorded it, you've narrated it, now edit it so it's clean and master it so it passes ACX Check.

Then share it with the world. Put it up on your blog, post it to your website, stick it in your newsletter. If this daunts you, simply email it to your mother or your best friend.

If you can do this, perhaps you can record an entire full-length novel as an audiobook.

As you go about your own recording experience, relax and enjoy yourself. You're creating beautiful art.

Let's get some stuff done!

The End

Backmatter

Afterword

Thank you for following on my journey for how I create an audiobook. I hope you found this information and accompanying exercises enlightening.

If you do choose to pursue creating your own audiobooks, I wish you well in your adventures and look forward to you creating a beautiful audiobook. May it delight thousands of listeners around the world.

About The Author

Author and Sound Engineer, Heidi Wessman Kneale

Ever since I was a child, I've been fascinated by books. Would read everything I could get my hands on. Loved them so much, I started writing my own. You'd think that with such a love, I'd move straight on into Creative Writing as a major at university.

I didn't. Instead, I studied Film and Music at the University of Utah in the United States in the early 90's (Class of 96!). The University of Utah is one of the original ARPANET universities, so computers were pretty much integrated into student life.

This was a fascinating time to be studying, as both film and sound engineering was going through a time of transition from analogue to digital. I got to produce sound pieces on both magnetic tape and on a Mac Classic computer. Because of my analogue training, I got into the habit early on of ensuring that the first track I laid down was the best I could do, because there was often no do-overs. (Ctrl-Z did not exist, and footage could be expensive.) Editing was a royal pain, and if you didn't set up your noise reduction and other effects before you laid down your recording, you couldn't fix it in post.

When computers came along, they were the bees-knees. It opened a whole new world with new opportunities for editing, correcting, post-production and more.

Applied an effect and didn't like how it sounded? Undo it and try again. You didn't have to worry about ruining your track. And if you did? No problem, because you'd already made a backup. (Okay, we did do backups in analogue, of a sort. We'd record our initial track, then make a copy. The original was our Master, and the copy is what we'd alter in post-production. Mess up this copy, and you had to make a new one from the Master and redo everything you'd already done. Yes, it was a pain.) When we get into the editing portion of this book, I'll deeply instil the age-old wisdom of making a backup of your raw WAV files before you start editing. I'm saying it here, because I cannot say it enough. And I'll say it again later. Backups.

I do not miss analogue post-production at all. AT ALL. Computers for the win.

I also studied some acting for the film side of stuff, had some voice training for the music side (I'm a mezzo-soprano coloratura, or rather, I was. My voice has smoked up a little bit with age) and had a few gigs on radio as a presenter for a local radio show you've never heard of, plus some voice-over work for various film projects. Since university, I've kept my skills up with regular live theatre gigs and frequent choral work. Occasionally I'd get an opportunity for some sound engineering for a roadshow or mixing music Because I Could.

Meanwhile, my time in Film School helped my author career more than any Creative Writing class ever did (and I did take lots of those in my Elective allotment). I met my writing mentor Dr Anne Wingate (of the Deb Ralston series fame) and she gave me that push I needed to get published.

Within a few weeks of graduating from university, I moved to Australia. Not much work in film or music here for me at the time, so I focused more on writing novels.

I watched as the publishing world went from paperback by traditional publishers to books, to the rise of independent publishing, to multiple formats, including audiobooks. Likewise, the 21st Century brought the kind of technology that meant sound production was no longer dependent on expensive, specialist equipment in the basement of some old church, but easily accessible on a simple computer. I got back into composing, arranging, mixing and more.

Then one day at a writer's conference, someone was telling me about the rise of audiobooks, and if I knew anything about it. Of course I knew about audiobooks, but production? The last time I had any exposure to creating what you could call an audiobook was back in the early 1990's. It involved a book, a quiet room and a tape recorder. Hardly professional-quality production, but students of the Utah School for the Blind didn't care. As long as they had access to the books they loved, it was all good.

Back to the 21st Century–I looked up what was involved in creating audiobooks with the new technology, and how one could go about getting them out there. By this time, ACX was a force in the market and Findaway Voices was holding strong. Because of this, there was no reason why I couldn't create audiobooks of my published titles and get them out there. After all, I had a sound engineering background and more than enough titles to make it work.

And that's how I came to produce my own audiobooks.

Since then, I've gotten back into talking for a living. Besides my own books, I've helped others bring their dreams to life, including some non-fiction work. (And if you call a certain educational institution looking for help in the technological area, the voice on the automated phone system happens to be mine...)

As long as I have a voice or can run a computer, I will continue to support the production of audiobooks. They are such a beautiful format. I've always loved being told stories. I grew up listening to Old Time Radio. I still go to sleep every night listening to someone telling me a story.

So, you want to turn your book into an audiobook? Sure, why not?

I did.

Resources

Listing all the resources I mention in this book so you don't have to go back and hunt them down. Note: this is not an exhaustive list of all possible resources found online that help an author become a better audiobook narrator. I encourage you to hit up your favourite search engine and explore further.

Websites

- My website where you can find links to all this information: https://heidikneale.com.au/AGRYOA
- Where to download the software editing application Audacity: https://www.audacityteam.org/download/
- Audacity's Audiobook Forum: https://forum.audacityteam.org/viewforum.php?f=64&sid=f9af0b4f0795c7b2f5ff4d12b7740ca8
- Here's the KoboBooks page for my audiobook "Her Endearing Young Charms" where you can listen to a sample https://www.kobo.com/au/en/audiobook/her-endearing-young-charms-2
- Australian Actor/Narrator Payscale information: https://www.payscale.com/research/AU/All_Actors_%2F_Actresses/Hourly_Rate
- Nathan Lowell's "The Golden Age of the Solar Clipper" series (science fiction), an audiobook narrated by the author https://scribl.com/books/P2A75/quarter-share

Glossary of Further Information

- Hertz: note frequencies and pitches https://en.wikipedia.org/wiki/Musical_note#Note_frequency_(hertz)
- Shaka sign from Hawaii https://en.wikipedia.org/wiki/Shaka_sign

- Singing pitches (human range) https://en.wikipedia.org/wiki/Vocal_range
- Decibels, the amplitude and loudness of sound https://en.wikipedia.org/wiki/Decibel
- Stereo sound https://en.wikipedia.org/wiki/Stereophonic_sound
- Mono sound https://en.wikipedia.org/wiki/Monaural_sound
- MP3 format for sound files (compact) https://en.wikipedia.org/wiki/MP3
- WAV format for sound files (lossless) https://en.wikipedia.org/wiki/WAV
- Royalty-free guidelines for sound use https://en.wikipedia.org/wiki/Royalty-free
- Public Domain https://en.wikipedia.org/wiki/Public_domain
- Public Domain info for music https://en.wikipedia.org/wiki/Public_domain#Public_domain_music
- Free Music https://en.wikipedia.org/wiki/Free_music

Narrator help online

- Patrick Fraley, who offers free narration lessons: https://patfraley.com/pf/free/
- Narrators Roadmap, an excellent all-round resource: http://www.narratorsroadmap.com/
- Tanya Eby asking if you want to be a narrator: http://www.tanyaeby.com/blog/2018/12/16/so-you-want-to-be-a-narrator
- Sean Pratt's test to see if you suit narration: https://www.youtube.com/watch?v=NPzPi-_0Xi8

Quality Submission Specs

- ACX: https://www.acx.com/help/acx-audio-submission-requirements/201456300
- Author's Republic: https://www.authorsrepublic.com/creation#spec
- INaudio: https://voices.inaudio.com/technical-requirements-for-assets

Music source sites

- Envato https://1.envato.market/eNQkO
- Envato's license terms: https://elements.envato.com/license-terms
- Media Music Now https://www.mediamusicnow.co.uk/royalty-free-music.aspx
- Media Music Now's license terms: https://www.mediamusicnow.co.uk/help-support/answers/what-can-your-music-be-used-for/audio-books.aspx
- Getty Images (TripleScoop Music) https://gettyimages.triplescoopmusic.com/
- Musopen https://musopen.org/

Audiobook Aggregators and Distributors

- Audible/ACX https://www.acx.com/
- Voices by INaudio https://www.inaudio.com/for-authors
- Author's Republic https://www.authorsrepublic.com/
- Kobo Writing Life https://www.kobo.com/us/en/p/writinglife
- Google Play Books https://play.google.com/books/publish/

Author organisations that offer contract review services

- ALLi – the Alliance of Independent Authors https://www.allianceindependentauthors.org
- The Australian Society of Authors https://www.asauthors.org.au/

YouTube Videos

- mine: How I Record an Audiobook: From Go to Whoa! https://www.youtube.com/watch?v=sFy8cDZ_XxU
- mine: Audiobook Narration Voice Techniques for Authors https://www.youtube.com/watch?v=c_DSH9vwclM
- Tuning the Orchestra https://youtu.be/npEz3h0h8CA
- Adam Neely explaining how slow is too slow: https://youtu.be/afhSDK5DJqA
- Sean Pratt's test to see if you suit narration https://www.youtube.com/watch?v=NPzPi-_0Xi8
- INaudio has a YouTube video for mastering in Adobe Audition: https://www.youtube.com/watch?v=x387JWPhKEg&t=452s)
- Stephen Fry on issues while narrating Harry Potter https://www.youtube.com/watch?v=ysu7DYRgE5o&t=1176s

Nyquist plug-in downloads for Audacity

- RMS-Normalise https://forum.audacityteam.org/t/rms-normalize/45334
- ACX Check https://wiki.audacityteam.org/w/images/5/53/Acx-check.ny
- Noise Gate https://plugins.audacityteam.org/Nyquist-plugins/effect-plugins/dynamics-processing

Download Files

These files are provided under a Creative Commons 4.0 License for the purpose of supporting the exercises and examples in this book. Due to the nature of some of these files, they are covered by one of two different license types:

CC BY-NC-ND (Attribution-NonCommercial-NoDerivatives). You may download these files for personal use. You must attribute them if you share them. You may not alter them, remix them, or create derivative works. Use of these files are only for non-commercial purposes.

CC BY-NC (Attribution-NonCommercial) You may download these files for personal use. You may alter them, remix them and create derivative works. You must attribute them if you share them. Use of these files are only for non-commercial purposes.

All these files are available to listen or download from the official webpage for "The Author's Guide to Recording Your Own Audiobooks" at https://heidikneale.com.au/AGRYOA/

CC license type	File name and location (URL)
BY NC	ch 1.3, p40 Words Per Minute https://heidikneale.com.au/wp-content/uploads/2026/07/words-per-minute.mp3
BY NC	ch 2.4, p64 Pride & Prejudice raw reading WAV file to download and play with: https://heidikneale.com.au/wp-content/uploads/2026/07/PP-raw-reading.wav MP3 for listening only: https://heidikneale.com.au/wp-content/uploads/2026/07/PP-raw-reading.mp3
BY NC	ch 2.4, p69 Pride and Prejudice – edited reading https://heidikneale.com.au/wp-content/uploads/2026/07/PP-edited-reading.mp3
BY NC ND	Ch2.4, p71 WHC Ambience 2 https://heidikneale.com.au/wp-content/uploads/2026/07/WHC-ambience-2.mp3
BY NC	ch 3.3, p87 Too Slow Speech https://heidikneale.com.au/wp-content/uploads/2026/07/tooslowspeech.mp3
BY NC	Ch3.3, p87 The Rainbow Passage at 140 wpm https://heidikneale.com.au/wp-content/uploads/2026/07/rainbowpassage_140.mp3 The Rainbow Passage at 160 wpm https://heidikneale.com.au/wp-content/uploads/2026/07/rainbowpassage_160.mp3

	ch 4.1, p112 Master_Sample_raw_track.wav https://heidikneale.com.au/wp-content/uploads/2026/07/Master_Sample_raw_track.wav
	Ch 4.1, p114 Master_Sample_mastered_track.mp3 https://heidikneale.com.au/wp-content/uploads/2026/07/Master_Sample_mastered_track.mp3
	ch4.3, p122 Chapter header sample.mp3 https://heidikneale.com.au/wp-content/uploads/2026/07/chapter-heading-sample.mp3
	ch4.3, p123, 000_Her_Endearing_Young_Charms.mp3 https://heidikneale.com.au/wp-content/uploads/2026/07/000_Her_Endearing_Young_Charms.mp3
	ch4.3, p124 019_Her_Endearing_Young_Charms_End_Credits.mp3 https://heidikneale.com.au/wp-content/uploads/2026/07/019_Her_Endearing_Young_Charms_End_Credits.mp3
	ch4.3, p135 Theme_music_HEYC.mp3 https://heidikneale.com.au/wp-content/uploads/2026/07/Theme_music_HEYC.mp3
	ch4.3, p135 Theme music: For Richer, For Poorer https://heidikneale.com.au/wp-content/uploads/2026/07/FRFP_theme_music.mp3
	ch4.3, p135 Incidental music: Merrybelle's Lament https://heidikneale.com.au/wp-content/uploads/2026/07/Merrybelles_Lament-2.mp3

Homework

A list and short summary of all homework assignments, designed to introduce you to the process of creating an audiobook. Lessons start with:

- observation of the ambient sound environment around you
- listening, observation and analysis
- learning how to use your equipment through recording
- narration experience, including analysis and critique
- gauging your speaking speed
- editing a short passage
- mastering a short passage
- combining all skills into a short project

For detailed information, please refer to the page listed in this book.

Page	Homework assignment summary
p26	Listen to the ambient sound in your environment. Are you aware of the sounds about you? Identify your current environment's "room tone".
p26	Get a recording device and record thirty seconds of ambient sound. Compare to your live experience.
p32	Record a minute or two of ambient sound (room tone), then do a voice check, recording several paragraphs of text. Listen to the recording, trying to identify any unwanted noise, including echoes, etc.
p33	Record you reading a few minutes of text and save it for later analysis. Then read section 2.1
p38	Go through the list of techniques in 2.1 and compare your previous recording. Listen to the

	recording with a critical ear and analyse it. Make notes of what you did right and what you would like to change.
p38	Make a timed five-minute recording of you reading a scene from your book. Do a word count from your book, then divide this by five to get your preliminary speaking speed. Also, listen with a critical ear. Did this performance improve compared to the previous one? Feel free to repeat this homework assignment. Practice will improve your narrating skills.
p69	Take a sound file you recorded. Listen through it and identify your unwanted noises: noisy breaths, coughs, sniffs, etc. Become aware of the unwanted noises.
p79	Grab your first recording of your voice (see p32). Analyse your voice. What is good about it? What would you improve?
p85	Read aloud this 100-word section from The Rainbow Passage and time yourself with a stopwatch. Use the table on p86 to determine your speaking speed. Do you need to speed up or slow down?
p96	Select a few different passages from your book, choosing different moods or vibes. Record yourself. Listen with a critiquing ear, analysing your performance.
p99	Grab an earlier recording of yours (p96) or lay down a fresh recording of a few paragraphs. Give it an editing pass (chapter 2.3) to clean it up. We'll be mastering it later (p113).
p112	Download this sample raw track and master it.
p113	Take that edited recording from p99 & apply the mastering steps. Can you sample pass ACX Check?
p129	Record a thirty-minute passage of text. Check it against the checklist.
p170	Take this thirty-minute passage of text that you recorded. Edit it, master it, and share it with the world.

Acknowledgements

I must thank the following who supported me for this book and other presentations on authors creating their own audiobooks:

Catherine Bilson, fellow narrator, audiobook producer and Romance author, who co-ran a panel with m at RWA to help other authors interested in audiobooks.

Romance Writers of Australia who ran my course "How to Record an Audiobook" for their RWA Academy.

All my students who came to learn how to create their own audiobooks and went away with so much knowledge and experience. Some of you have created some really nice audiobooks. Glad you've been successful in your endeavours.

Index

www.ingramcontent.com/pod-product-compliance
Lightning Source LLC
LaVergne TN
LVHW050933210926
848482LV00013B/602
* 9 7 8 1 9 2 3 8 5 1 0 0 9 *